Sustainability in Latin America

Sergio Gabriel Ceballos Pérez
Carlos Alberto Zuniga González
Alejandro Téllez Jurado • Yuridia Mercado Flores
Editors

Sustainability in Latin America

A Bioeconomic Approach

Springer

Editors
Sergio Gabriel Ceballos Pérez
Universidad Politécnica de Pachuca
Zempoala, Hidalgo, Mexico

Carlos Alberto Zuniga González
Uni. Nacional Autónoma de Nicaragua
Leon, Nicaragua

Alejandro Téllez Jurado
Universidad Politécnica de Pachuca
Zempoala, Hidalgo, Mexico

Yuridia Mercado Flores
Universidad Politécnica de Pachuca
Zempoala, Hidalgo, Mexico

ISBN 978-3-032-09786-6 ISBN 978-3-032-09787-3 (eBook)
https://doi.org/10.1007/978-3-032-09787-3

This Springer imprint is published by the registered company Springer Nature Switzerland AG
The registered company address is: Gewerbestrasse 11, 6330 Cham, Switzerland

Contents

Chapter 1
The Role of the Bioeconomy in Achieving Sustainability

Sergio Gabriel Ceballos Pérez, Carlos Alberto Zuniga González, Alejandro Téllez Jurado, and Yuridia Mercado Flores

Abstract Bioeconomy plays a key role in achieving sustainability, as it seeks to harness natural resources, biodiversity, and biological technologies to reduce pollution, promote economic growth, and improve social well-being equitably. Bioeconomy can reduce greenhouse gas emissions and improve the integration of the circular economy by transforming organic waste and biomass into valuable products. Likewise, bioeconomy drives innovation, job creation, biodiversity conservation, and social well-being by creating networks within communities. Bioeconomy can contribute to food security and rural development through agricultural practices that improve soil health and crop yields and provide new sources of income for communities. It also provides benefits to human health and the environment through the development of pharmaceutical products. This book presents various theoretical and applied studies of the bioeconomy in Latin America from a multidisciplinary perspective and methods, encompassing economics, biotechnology, sociology, and others.

Keywords Sustainable · Circular economy · Bioeconomy · Latin America

1.1 Some Contributions from Latin America to the Bioeconomy

Sustainability has become an essential goal for contemporary societies, facing the pressing challenges of climate change, environmental degradation, and the depletion of natural resources. However, one of the main challenges facing sustainability is the

S. G. Ceballos Pérez (✉) · A. Téllez Jurado · Y. Mercado Flores
Laboratorio de Aprovechamiento integral de Recursos Bióticos, Universidad, Politécnica de Pachuca, Carr. Pachuca-Cd-Sahagún, Zempoala, Hidalgo, México
e-mail: sgceballospe@secihti.mx; alito@upp.edu.mx; yuridiamercado@upp.edu.mx

C. A. Zuniga González
Zúniga-González Carlos Alberto, edificio central, UNAN León, León, Nicaragua
e-mail: czuniga@ct.unanleon.edu.ni

S. G. Ceballos Pérez et al. (eds.), *Sustainability in Latin America*,
https://doi.org/10.1007/978-3-032-09787-3_1

need for more promising results, or put another way, the ambiguity with which the concept is understood. Bioeconomy, which encompasses the production and utilization of biological resources, knowledge, science, and technology, offers a promising pathway to achieve sustainability (Calicioglu & Bogdanski, 2021). By replacing fossil-based resources with renewable biological ones, the bioeconomy can contribute to the ecological, economic, and social pillars of sustainability (Bugge et al., 2016).

One of the most significant ecological benefits of bioeconomy is its potential to reduce greenhouse gas emissions. By shifting from fossil fuels to bio-based energy sources, such as biofuels and biomass, carbon dioxide emissions can be notably decreased (Tan & Lamers, 2021). For instance, biofuels produced from agricultural residues or algae can serve as sustainable alternatives to gasoline and diesel, significantly lowering the carbon footprint of the transportation sector (Fernández et al., 2021; D'Amato et al., 2020).

The bioeconomy promotes the principles of a circular economy by transforming waste materials into valuable products (Kardung et al., 2021). Organic waste from agriculture, forestry, and household activities can be processed into biogas, compost, or biochar, thus reducing landfill waste and minimizing environmental pollution (D'Amato et al., 2020). This approach not only conserves resources but also creates new economic opportunities and reduces the ecological impact of waste disposal (Skrivervik, 2020).

Sustainable bioeconomic practices can support biodiversity by encouraging the preservation and sustainable management of natural habitats (Bastos Lima & Palme, 2022). Agroforestry, for example, integrates trees and shrubs into agricultural landscapes, enhancing habitat diversity and promoting wildlife conservation (Thorenz et al., 2018). Additionally, sustainable fishing and aquaculture practices can help maintain healthy marine ecosystems, ensuring the long-term viability of fish populations and marine biodiversity (Gicana et al., 2022).

The transition to a bioeconomy has the potential to generate a substantial number of green jobs across various sectors, including agriculture, forestry, biotechnology, and renewable energy (Dietz et al., 2018). These jobs not only contribute to economic growth but also promote social inclusion and provide new opportunities for rural and urban communities alike (Calicioglu & Bogdanski, 2021). By investing in education and training programs, societies can equip their workforce with the necessary skills to thrive in a bio-based economy (Kardung et al., 2021).

The bioeconomy drives innovation by fostering the development of new technologies, products, and processes (Bröring et al., 2020). Advances in biotechnology, for instance, enable the creation of bio-based materials, such as bioplastics and bio-based chemicals, which can replace their fossil-derived counterparts (Wilde & Hermans, 2021). This innovation enhances the competitiveness of industries by opening new markets and reducing dependency on non-renewable resources (Kuckertz et al., 2020). As a result, countries that invest in bioeconomy can position themselves as leaders in sustainable development (Stegmann et al., 2020).

A well-functioning bioeconomy can contribute to food security by promoting sustainable agricultural practices and increasing the resilience of food systems

(Calicioglu & Bogdanski, 2021). Techniques such as crop rotation, organic farming, and precision agriculture can enhance soil health, increase crop yields, and reduce the reliance on chemical inputs. Furthermore, the bioeconomy can stimulate rural development by providing new income sources for farmers and rural communities, reducing poverty, and improving overall quality of life (Johnson & Altman, 2014; Papadopoulou et al., 2021).

The bioeconomy also has the potential to improve public health and well-being. For instance, the development of bio-based pharmaceuticals and nutraceuticals can lead to innovative treatments and preventive measures for human and animal diseases (Sadhukhan et al., 2024). Additionally, reducing pollution through sustainable waste management and utilizing alternative energy sources can contribute to improved air and water quality, ultimately benefiting both ecosystems and human health (Tufail et al., 2022).

The bioeconomy presents a multifaceted approach to achieving sustainability by addressing ecological, economic, and social challenges (Bugge et al., 2016). By harnessing the power of biological resources and technologies, societies can reduce their dependence on fossil fuels, create new economic opportunities, and enhance the well-being of their populations (Kardung et al., 2021). As the world continues to grapple with the urgent need for sustainable development, the bioeconomy offers a viable and promising path forward (Tan & Lamers, 2021). Investing in bioeconomic innovations and practices is not only a strategic choice but also a necessary step toward a sustainable and resilient future.

This book addresses issues related to bioeconomy as a tool for transforming the production and consumption focus, contributing to sustainable development. Chapters highlight the importance of sustainable economic growth through the bioeconomy with minimal environmental impact on natural resources and climate change, while focusing on the welfare of the inhabitants of Latin America. It is emphasized that the use and development of science and technology, in combination with traditional and community knowledge, facilitate the transition toward sustainable practices and the circular economy.

Chapter 2 begins with a review of the principles of bioeconomy and its development through the theory of energy value proposed by Georgescu-Roegen, whose biophysical basis of thermodynamic theory and the Second Law of Entropy is linked to epistemological currents in Latin America and the Caribbean (LAC) to explain the emergence of a new economic paradigm called "Latin Bioeconomy", which includes three new concepts called *Bioculture, Bioterritoriality*, and *Bioinformation.* The chapter proposes a new model that integrates Latin American political philosophies of well-being, such as living well, with orthodox economics, complementing it and making it multidisciplinary, thus expanding its theoretical and methodological coverage to address problems currently affecting humanity.

Global initiatives focus on creating economic valorization chains that encourage the efficient use of renewable resources and the reduction of pollutants, promoting the valorization of agro-industrial residual. For this, the biorefinery and life cycle analysis are tools for designing the processes and products to be integrated into the new economy. Chapter 3, "Life Cycle Analysis as a Tool for the Evaluation of the Environmental Impact of a Coffee Residual Mixture Biorefinery," offers innovative approaches to integrate sustainability. The coffee industry generates a considerable amount of coffee products worldwide. Because of its industrial processing, a large volume of residual or by-products is generated in the various processing stages; said residue is considered highly polluting and is rarely disposed of responsibly. A biorefinery allows waste to be used by converting it into new, high-value products that are useful for other industries or for generating energy, while significantly reducing pollution.

Agro-industrial residues, including leaves, roots, straw, bagasse, and husks, are abundant products of human activities like agriculture and livestock. In many Latin American countries, these wastes are not adequately regulated for treatment before disposal. Although waste management strategies have been proposed, their implementation faces obstacles, often leading to harmful practices such as burning or landfilling, which pose economic, environmental, and social issues. Chapter 4 addresses issues of valorization of agro-industrial waste using green technologies, taking as an example the waste generated from the processing of green tomatoes, pumpkin residue, and apple bagasse in Mexico, it is proposed to incorporate them into a circular economy into a circular economy model through the generation of by-products of food, pharmaceutical and cosmetic interest using these residues as raw material.

Another important challenge that the bioeconomy can solve is the treatment of contaminated water from slaughterhouses, which generate large volumes of wastewater because of the slaughter and processing of bovine, porcine, ovine, caprine, equine, and poultry animals. This wastewater is characterized by a high pollutant load, including chemical oxygen demand (COD), total suspended solids (TSS), fats, and nutrients such as phosphorus and nitrogen. The direct discharge of this untreated water into surface water bodies has caused eutrophication, ecosystem degradation, and serious public health risks due to the proliferation of pathogenic microorganisms. Chapter 5, titled “Treatment of Slaughterhouse Wastewater for Reuse and Waste Utilization as a Circular Economy Alternative,” discusses the potential for utilizing waste produced from slaughterhouse wastewater treatment in various bioprocesses. This approach aims to contribute to agriculture, bioenergy, and the production of bioproducts, all viewed through the lens of a circular economy.

Revisiting the Latin American bioeconomy model from Chap. 2, in which bioculture is one of the variables introduced, Chapter 6 highlights the importance of biocultural values in agribusiness and tourism, demonstrating how cultural diversity can enhance sustainability in different territories. This chapter offers a platform for reflecting on the revaluation of local biocultural resources, such as the ahuilote (*Vitex mollis* Kunth). This species can serve as a model to broaden the understanding of the connection between the economy and nature in regions transitioning toward a utilitarian model of natural resources. Despite being ignored in certain areas, this fruit of pre-Hispanic heritage represents an opportunity in the Valles de Jalisco region of Mexico for the utilization and integration of cultural knowledge with a sustainable approach to the use of natural resources.

Extensive cultivation, like cotton, causes significant alterations in the local ecosystems due to the excessive use of pesticides. This excess allows pests to develop resistance to these products, resulting in the resurgence of secondary pests that previously had no economic relevance. Among these, the bollworm (*Helicoverpa zea*), cotton leafworm (*Alabama argillacea*), and various species belonging to the genus *Spodoptera* stand out. Effective management of agricultural pests using biological controllers, such as *Sitotroga cerealella*, *Trichogramma*, and *Chrysoperla*, is crucial for sustainable pest control strategies. An in-depth cost analysis of these organisms is essential to optimize production processes and establish competitive pricing. A comprehensive understanding of these cost dynamics facilitates better financial planning and supports the development of effective pricing strategies for biological pest control products.

Another characteristic of the Latin American bioeconomy is the cooperation that must be fostered between communities, universities, governments, and businesses. Chapter 8, entitled "Composting Training Program: A Collaborative Response That Contributes to the Circular Bioeconomy of Organic Solid Waste," analyzes the joint efforts of the Xalapa City Council and the University of Veracruz to implement a virtual composting training program. This program, aimed at the general population, had positive results, as participants were able to produce compost. This initiative demonstrates how collaboration between government and academic institutions can strengthen community skills essential for effective organic waste management.

Chapter 9 "Sustainable Innovation: Nanocellulose Extraction with Natural Deep Eutectic Solvents in Lignocellulosic Residues in Latin America" focuses on sustainable innovation, taking the production of nanocellulose as an example, nanocellulose using natural eutectic solvents and lignocellulosic waste generated in Latin America. The potential of lignocellulose as a raw material for high-value compounds obtained from green technologies, such as deep eutectic solvents, is exposed.

Latin America has been driving the advancement of the bioeconomy with a commitment to substitute fossil fuels with renewable energies in response to the global environmental crisis. Chapter 10, titled "Analysis of Agroecological Bioeconomy," focuses on a bibliographic analysis that highlights the emergence of the term agroecological bioeconomy as a new field of study that demands an integrated research approach with practical applications aimed at advancing sustainable development in agriculture and its related components. An "agroecological bioeconomic" perspective values the alternative of those who argue that agro industrial monocultures make farmers dependent on agrochemicals. External contributions undermine and distance the consumer's knowledge base. In Latin America, proponents of this viewpoint come from both the public and private sectors. Organic agriculture research institutes and movements to support them including the organic department of specialized agricultural organizations, environmental NGOs.

Finally, Chap. 11, titled " Exergetic Scope of Bioeconomics: The Sustainable Energy Transition ", examines the environmental impact of Mexican oil compared to renewable energy, emphasizing the potential of renewable technologies to optimize biodiversity exploitation. It focuses on the relevance of the threat posed by climate change, an emergency primarily derived from the effects of fossil fuel combustion. It argues that, under existing technological conditions in Mexico, the gap between the energy density of fossil fuels and the potential energy of renewable technologies—the exergy of alternative technologies—in addition to the energy policy shift of the last seven years makes the transition to a predominance of renewable energy an unlikely scenario even in these times. Given the energy requirements demanded by population and economic growth, greater adjustments and investments are required, especially in infrastructure and transition technology. For example, bioenergy has significant untapped potential but should be used in circular models for agroindustry, urban areas, etc.

References

Bastos Lima, M. G., & Palme, U. (2022). The bioeconomy–biodiversity nexus: Enhancing or undermining nature's contributions to people? *Conservation, 2*(1), 7–25. https://doi.org/10.3390/conservation2010002

Bröring, S., Laibach, N., & Wustmans, M. (2020). Innovation types in the bioeconomy. *Journal of Cleaner Production, 266*. https://doi.org/10.1016/j.jclepro.2020.121939

Bugge, M. M., Hansen, T., & Klitkou, A. (2016, July 19). What is the bioeconomy? A review of the literature. *Sustainability (Switzerland)*. MDPI. https://doi.org/10.3390/su8070691

Calicioglu, Ö., & Bogdanski, A. (2021). Linking the bioeconomy to the 2030 sustainable development agenda: Can SDG indicators be used to monitor progress towards a sustainable bioeconomy? *New Biotechnology, 61*, 40–49. https://doi.org/10.1016/j.nbt.2020.10.010

D'Amato, D., Veijonaho, S., & Toppinen, A. (2020). Towards sustainability? Forest-based circular bioeconomy business models in Finnish SMEs. *Forest Policy and Economics, 110*. https://doi.org/10.1016/j.forpol.2018.12.004

Dietz, T., Börner, J., Förster, J. J., & von Braun, J. (2018). Governance of the bioeconomy: A global comparative study of national bioeconomy strategies. *Sustainability (Switzerland), 10*(9). https://doi.org/10.3390/su10093190

Fernández, F. G. A., Reis, A., Wijffels, R. H., Barbosa, M., Verdelho, V., & Llamas, B. (2021). The role of microalgae in the bioeconomy. *New Biotechnology, 61*, 99–107. https://doi.org/10.1016/j.nbt.2020.11.011

Gicana, R. G., Yeh, F. I., Hsiao, T. H., Chiang, Y. R., Yan, J. S., & Wang, P. H. (2022). Valorization of fish waste and sugarcane bagasse for Alcalase production by bacillus megaterium via a circular bioeconomy model. *Journal of the Taiwan Institute of Chemical Engineers, 135*. https://doi.org/10.1016/j.jtice.2022.104358

Johnson, T. G., & Altman, I. (2014). Rural development opportunities in the bioeconomy. *Biomass and Bioenergy, 63*, 341–344. https://doi.org/10.1016/j.biombioe.2014.01.028

Kardung, M., Cingiz, K., Costenoble, O., Delahaye, R., Heijman, W., Lovrić, M., et al. (2021). Development of the circular bioeconomy: Drivers and indicators. *Sustainability (Switzerland), 13*(1), 1–24. https://doi.org/10.3390/su13010413

Kuckertz, A., Berger, E. S. C., & Brändle, L. (2020). Entrepreneurship and the sustainable bioeconomy transformation. *Environmental Innovation and Societal Transitions, 37*, 332–344. https://doi.org/10.1016/j.eist.2020.10.003

Papadopoulou, C. I., Loizou, E., Melfou, K., & Chatzitheodoridis, F. (2021). The knowledge based agricultural bioeconomy: A bibliometric network analysis. *Energies*. MDPI. https://doi.org/10.3390/en14206823

Sadhukhan, J., Martinez-Hernandez, E., Amezcua Allieri, M. A., Zermeño Eguía-Lis, J. A., Castillo, A., Dominguillo, D., et al. (2024). Strategic navigation of world-leading biorefineries and Mexico's policy landscape: A gateway to a sustainable circular bioeconomy. *Journal of Cleaner Production*. Elsevier Ltd. https://doi.org/10.1016/j.jclepro.2023.140386

Skrivervik, E. (2020). Insects' contribution to the bioeconomy and the reduction of food waste. *Heliyon, 6*(5). https://doi.org/10.1016/j.heliyon.2020.e03934

Stegmann, P., Londo, M., & Junginger, M. (2020). The circular bioeconomy: Its elements and role in European bioeconomy clusters. *Resources, Conservation and Recycling: X*. Elsevier B.V. https://doi.org/10.1016/j.rcrx.2019.100029

Tan, E. C. D., & Lamers, P. (2021). Circular bioeconomy concepts—A perspective. *Frontiers in Sustainability, 2*. https://doi.org/10.3389/frsus.2021.701509

Thorenz, A., Wietschel, L., Stindt, D., & Tuma, A. (2018). Assessment of agroforestry residue potentials for the bioeconomy in the European Union. *Journal of Cleaner Production, 176*, 348–359. https://doi.org/10.1016/j.jclepro.2017.12.143

Tufail, T., Ain, H. B. U., Saeed, F., Nasir, M., Basharat, S., Mahwish, et al. (2022). A retrospective on the innovative sustainable valorization of cereal bran in the context of circular bioeconomy innovations. *Sustainability* (Switzerland). MDPI. https://doi.org/10.3390/su142114597

Wilde, K., & Hermans, F. (2021). Innovation in the bioeconomy: Perspectives of entrepreneurs on relevant framework conditions. *Journal of Cleaner Production, 314*. https://doi.org/10.1016/j.jclepro.2021.127979

Chapter 2
The Bioeconomic Bridge Between Georgescu-Roegen and the Latino Bioeconomy

Carlos Ricardo Menéndez-Gámiz, Jorge Quiroga-Canaviri, Carlos Alberto Zuniga González, and Sergio Gabriel Ceballos Pérez

Abstract Background: The purpose of this research is to study the bioeconomy of the renowned Ecological Economist and father of the European Bioeconomy Nicholas Georgescu-Roegen, which is linked to the epistemological currents in Latin America and the Caribbean (LAC) to explain the emergence of a new economic paradigm called "Latin Bioeconomy" that includes the theory of energy value proposed by Georgescu-Roegen, whose physical basis of the thermodynamic theory and the second Law of Entropy is enriched with three new concepts called *Bioculture*, *Bioterritoriality*, and *Bioinformation*. Methodology: The research was of an exploratory documentary using VosViews bibliometric tools to explain how one production model is developed from another. The open innovation method is used that articulates the convergence of ancestral and modern knowledge, translated into bioculture, which coincides with the bibliographic review of authors identified as epistemologists of the South, who disagree with the absolutist models of orthodox economies. The deductive method is also followed to deduce conclusions that go from the general to the particular. Results and discussion: A space is created for Latin American political philosophies of Welfare such as Vivir Bien/Vivir Bien, to give social content to orthodox economics, complementing it, making it multidisciplinary, and expanding its theoretical-methodological coverage, to solve problems that currently affect humanity. The discovery is defined as an

C. R. Menéndez-Gámiz
Agro-Industrial Economic Problems, Chapingo Autonomous University, Texcoco, Mexico

J. Quiroga-Canaviri
Economics, Mayor University of San Andrés (UMSA), Santa Cruz, Bolivia

C. A. Zuniga González (✉)
Agroecology and Agri-Business, Universidad Nacional Autonoma De Nicaragua, Leon, Nicaragua
e-mail: czuniga@ct.unanleon.edu.ni

S. G. Ceballos Pérez
Dirección de Investigación, Innovación y Posgrado, SECIHTI-Universidad Politécnica de Pachuca, Zempoala, Mexico
e-mail: ceballosp@upp.edu.mx

S. G. Ceballos Pérez et al. (eds.), *Sustainability in Latin America*,
https://doi.org/10.1007/978-3-032-09787-3_2

epistemology of Latin American sentiment that provides answers to the European energy theory of value proposed by the Romanian Georgescu-Roegen, also adding the concepts of bioculture, bioterritoriality, and bioinformation.

Keywords Bioeconomy · Latin America · Bioculture · Bioterritoriality · Bioinformation

2.1 Introduction

Georgescu-Roegen noted: "Pindar, the poet who lived a hundred years before Plato, was aware of the famous paradox of value, where expensive jewels are valued despite lacking utility, and of the valuable air that is a free good" (Georgescu-Roegen, 1983, p. 289). The paradox describes scarcity and value from the historical perspective of several philosophers who for centuries attempted to attribute economic value to a single cause, promoting "monism" as the dominant guide in their scientific investigations, a situation this manuscript avoids Georgescu-Roegen (1974), Georgescu-Roegen (1979), Georgescu-Roegen (1990).

Although Western culture elevated the values of freedom, equality, and property, these were not transferred in the same condition to Latin America. Due to the notable divergences, the European invaders of the Americas (referred to by indigenous cultures as Abya Yala[1]) imposed their philosophical roots on the Andean, Amazonian, and Mesoamerican mother cultures, managing to interrupt the process of peace and restructuring of matter, energy, which should be fluid and restored to recover the balance between human beings and nature (Gudynas and Acosta, 2011; Quiroga & Zaiduni, 2021).

This is where the Georgescu-Roegen bioeconomy fits in, incorporating low-entropy energy through the laws of thermodynamics into classical and linear production functions (Georgescu-Roegen, 1971, 1975, 1983). This becomes the key to elucidating the dilemma indicated. However, based on previous research, three ingredients were missing to fully address the three main problems that he and other scientists predicted 53 years ago (Meadows et al., 1972, 2012): climate change, food insecurity, and pandemics, which humanity itself caused due to its selfishness, dominated by its rebellion against nature, of which it is also a part.

The three missing ingredients are *bioculture*, *bioterritoriality*, and *bioinformation*, which many economists and social scientists have omitted for centuries (Gómez Restrepo, 2024; McDonough & Braungart, 2003; Ceballos, 2019; Martínez, 2009). These are complemented by multidimensionality and multisectorality, which Georgescu-Roegen energetically introduced, and linked to

[1] The term "Abya Yala" comes from the Kuna language and means "mature land" or "land in full maturity" and is used by the indigenous peoples of the Americas to refer to the continent. This term is a way of rejecting the name imposed by the European colonizers and of reclaiming the identity and history of indigenous peoples.

the philosophy of well-being (Sen, 2014) from the Latin American perspective, understood institutionally as Living Well or Good Living in Bolivia and Ecuador, respectively, as well as the institutional bioeconomy strategies of Colombia, Argentina, Mexico, Costa Rica (Chavarría, 2022), and the non-institutional but rather popular bioeconomy driven by the actors (Ceballos Pérez & Azamar Alonso 2022).

Thanks to Georgescu-Roegen, the bioeconomy is considered the best nature-friendly production paradigm, capable of reincorporating the residual biomass emerging from the recycling of waste produced by agriculture and other production processes that generate biofuels (substitutes for fossil fuels), soil bioremediators, bioinputs, biomedical products, nutraceuticals, biocosmetics, bioservices (agro-bioecological tourism), and a wide range of bioproducts and bio-products. Bioservices, while meeting human needs, also protect the biosphere of planet Earth (Scarlat et al., 2015; Georgescu-Roegen, 1977, 1975).

What stands out today is the reappearance of consumerist and utilitarian linear practices typical of classical orthodoxy in circular bioeconomic environments, which requires a careful review of Georgescu-Roegen, whose mathematical and physical arguments are once again gaining relevance (Zúniga et al., 2022). However, ancestral Latin cultures were respectful of nature and the processes of space-time and matter-energy restructuring, elements that classical and neoclassical orthodox models simplified or omitted, marginalizing the underlying physics (Forero et al., 2006). It was confirmed that there are no absolute truths, and the euphemisms of objective orthodoxy, so it would be worth reflecting before rejecting the proposals of new epistemologies.[2,3]

Finally, the chapter concludes by showing the convergence of the approaches of Georgescu-Roegen and Latin American Bioeconomics, both irreverent to orthodoxy, daring to propose new routes or paradigms that are proposed as a guiding factor for harmonious productive development adjusted to reality in other economies of developing countries, where classical macroeconomics (fiscal and monetary theories and policies) and classical microeconomics (utility, theories of individual production and consumption) are in a steady state in the face of the ongoing climate crisis, persistent food insecurity exacerbated by wars, and the re-emergence of COVID-19 variants that continue to be a burden on humanity (Zúñiga-González, 2023).

---

[2]*Arithmetic aspects* refer to the mathematical and statistical methods applied to understand economic phenomena, and *ceteris paribus* refers to the condition in which all other variables, except capital and labor, are constant.

[3]The year 2006 marks the centenary of the birth of Nicholas Georgescu-Roegen (1906–1994). Óscar Carpintero published one of the few works in Spanish (Spanish) about the Romanian economist. The fact that he published his work in Spanish, "a language of limited scientific use" (Martínez-Alier et al., 2008, p. 148) considered as such by conventional epistemology, was an intellectual audacity, but the profound content revealed that Oscar Carpintero's "audacity" was much more substantial than classical epistemological approaches.

2.2 Methodology

The research carried out is of an exploratory documentary type, identifying in various research platforms the bibliography produced by different experts in bioeconomy; both supporters and critics of Georgescu-Roegen's work found through keywords and bibliometric tools that allow the analysis and quantification of scientific production in post-Georgescu-Roegen bioeconomy. Indicators such as the number of publications, citations, authors, and indexed journals were used to identify and organize relevant information (Zhu & Liu, 2020). Keywords are specific terms that represent concepts inherent to the research topic, facilitating the retrieval of relevant documents from bibliographic databases, many of which were used and cited in this manuscript Georgescu-Roegen (1990), Georgescu-Roegen (1994).

2.2.1 Data Sources

For the preparation of this article, fundamental data were extracted from the two main widely recognized databases: WoS and Scopus (Zhu & Liu, 2020). Given the growing relevance of new trends in research, the inclusion of both sources enriches the perspective on emerging directions in a specific research area (Echchakoui, 2020). Consequently, bibliometric analysis involves a process of reviewing and excluding duplicate documents, which is essential to ensure high consistency between WoS and Scopus (Aksnes & Sivertsen, 2019) (Tables 2.1 and 2.2).

2.2.2 Analysis Tool

For the literature review, the VOSviewer software (RRID: SCR_023516) was used using the keywords "Bioeconomy (Bioeconomy) and/or Bio-Economy." A total of 6037 documents were analyzed; of these, 4905 are secondary or non-indexed

Table 2.1 Search criteria

Variables	Results
Database	WOS, Scopus
Time frame	2005–2024; North and South America
Document type	12 January 2024
Type of journal	All types
Search field	Title
Search terms	Bioeconomy or Bio-Economy
Language	English
Results	331–4905
Total results	6037

Source: Authors, Scopus and WOS databases

Table 2.2 Ten authors with the most articles and citations in WoS and Scopus (2005-January 2024)

No	Author's name	Documents by author	Affiliation	Documents by affiliation
1	Blumberga, D.	61	Technical University of Riga	62
2	Thrän, D.	54	Helmholtz Centre for Environmental Research	56
3	Pandey, A.	40	German Non-Profit Biomass Research Center GmbH	47
4	Thrän, D.	35	Indian Institute of Toxicological Research	36
5	Varjani, S.	33	Institute for the Study of Nanostructured Materials, Rome	34
6	Lewandowski, I.	28	National Research Council	34
7	Toppinen, A.	27	Gujarat Pollution Control Board	31
8	Ciriminna, R.	27	University of Hohenheim	27
9	Muizniece, I.	25	University of Helsinki	27
10	Santi, T.	4	Special to O Papel, Brazil	1

Source: Authors, Scopus and WoS databases 2015–2024

documents but are cited by indexed documents on the topic of bioeconomy (Zúñiga-González, 2023; Huang et al., 2020; Van Eck & Waltman, 2010).

A descriptive method is used to study and understand the European bioeconomy in depth to develop the Latin American bioeconomy model based on the work of Georgescu-Roegen. This method is complemented by the open innovation method (Ramírez-Montoya & García-Peñalvo, 2018), which consists of examining the key elements of the existing model. First, a detailed description of the structure, processes, and components of the original model was provided, and then potential innovations resulting from the convergence of ancestral and modern knowledge and other omissions were identified. The open innovation method is useful for seeking and adopting ideas, technologies, or practices external to the Georgescu-Roegen bioeconomic system. The combination of the descriptive approach with open innovation helps to fully understand the current model, identify its limitations, and explore innovative solutions from internal and external sources. This facilitates the development of a new production model adapted to the current needs and the different realities of the countries of the Global South.

2.3 The Epistemological Encounter: The Latin American Bioeconomy from the Bioeconomy of Georgescu-Roegen

Among the most notable aspects of the legacy of Nicholas Georgescu-Roegen, who died in 1994, are books and essays considered fundamental to the development of the Latin American bioeconomy. These include the following: "The Law of Entropy and the Economic Process" (1971), "Dynamic Models and Economic Growth" (1974), "Energy and Economic Myths" (1975), "What Can Thermodynamics and Biology Teach Economists?" (1977), "Energy Analysis and Economic Valuation" (1979), "The Energy Theory of Economic Value: A Particular Economic Sophistry" (1983), "Thermodynamics, Economics and Information" (1990), and "Nicholas Georgescu-Roegen on Himself" (1992).

The central aspects of Georgescu-Roegen's mathematical and critical work, which led him to move from economic orthodoxy to heterodox economics, questioned the most solid foundations of classical and neoclassical theories of consumption, production, and income distribution. According to Carpintero (2006), the mathematical inconsistencies of the assumptions, the arithmorphic aspects, the standardizing nature of homo economicus, the ceteris paribus assumption, and the ahistorical and ageographic character of the models undermined the conventional functions.

For example, homogeneous functions such as the Cobb-Douglas function $Q = A{\cdot}K\ \alpha{\cdot}L$ where $\alpha + \beta = 1$ were replaced by complex nonlinear and multivariate (multisectoral) functions that underpin the bioeconomy (Eq. 2.1).

$$Q_0(t) = F\begin{bmatrix} T & T \\ I_i(t); & W_o(t) \\ 0 & 0 \end{bmatrix} \tag{2.1}$$

For Georgescu-Roegen (1971), inputs (I) and waste or garbage (W) are involved in the production process, where $i = 1... ni = 1\ldots ni = 1...n$ (resources range from 1 to n, including nature, biodiversity, and others). The introduction of the residual variable in the model focuses on incorporating the laws of thermodynamics, considering that human economic activity, as a characteristic of biological evolution, can be described in thermodynamic terms and through its second law as a nonlinear function of low entropy. This function underscores the importance of considering the depletion of natural resources and other fundamental elements in the production function.

The function goes from 0 to T as the service funds flow. It recovers the functions of each of the factors in a circular way according to their physical qualities. In order for the model to work in a steady state, it proposes to maintain the efficiency of durable factors of production (Hernández-Cervantes, 2008, p. 40).

In 1974, Georgescu-Roegen stylized the production function (Eq. 2.2) as follows:

$$Q_0(t) = F\begin{bmatrix} T & T & T & T & T & T & T & T \\ R(t) & I(t) & M(t) & Q & W(t) & L(t) & K(t) & H(t) \\ 0 & 0 & 0 & 0 & 0 & 0 & 0 & 0 \end{bmatrix} \tag{2.2}$$

The second part of the equation, after the colon, groups the stock market variables: Labor (L), Physical Capital (K), and Ricardian Land (H). The first part of the equation groups the flow variables: Natural Resources (R) such as rain, solar energy, soil, etc.; Current Inputs or materials derived from other production processes (I), such as varnish for furniture, and other materials (M) necessary to keep capital goods intact, such as lubricants, paints, parts and others; and the production function (Q) itself as the desired result of the process.

Equation (2.2) is a function of functions, temporally limited within the space from 0 to T, where several elementary processes occur. Within the 0 to T space, several elementary processes can occur within a range of 0 to t, where $t < Tt < T$ always occur (Hernández-Cervantes, 2008).

It is here that Georgescu-Roegen stopped, incorporating previous reflections on agricultural economics (Georgescu-Roegen, 1967), positioning himself as a dissident against the dominant universal economic approaches of his time (neoclassical and Marxist), due to their null applicability in Latin American economies, raising debates and differences between agricultural and industrial production processes. In addition, according to Carpintero (2006, p. 148), Georgescu-Roegen, in order to navigate the turbulent waters of interdisciplinarity, relied heavily on the book by the physicist Erwin Schrödinger, *What Is Life?*, emphasizing that life was not alien to the laws of thermodynamics. This led to the publication of *Analytical Economics* in 1966, which compiled several works that established the argumentative threads of the emerging bioeconomic approach (Georgescu-Roegen, 1966) and underpin Eq. (2.2) presented in this manuscript.

However, considering Georgescu-Roegen's work alone was not enough. It was necessary to understand contemporary advances in bioeconomy supported by the bibliometric analysis of all scientific contributions on bioeconomy, as well as bioeconomic models adapted to Latin America (Zuniga-Gonzalez et al., 2024) and emerging currents (Ordoñez & Lakner, 2023), as pointed out (Bugge et al., 2016), which served as the analytical basis for a previous essay (Quiroga-Canaviri & Menéndez-Gámiz, 2023).

The present manuscript expands upon and systematically updates the bibliometric information and analysis provided by previous studies (Perea Mosquera et al., 2021; Alzate Cárdenas et al., 2022; Mougenot & Doussoulin, 2022; and Gould et al., 2023) to reflect current advances in the bioeconomy. Specifically, Table 2.3 updates the references for the best-ranked authors—originally cataloged in the bibliometric reports by Alzate Cárdenas et al. (2022) and Gould et al. (2023)—with current citations through January 2024.

As you can see, this list does not include Latino authors, although there are many in Brazil, Mexico, Argentina, Colombia, Nicaragua, Costa Rica, and other countries in Central and South America. Therefore, another bibliometric analysis was performed, which presents the results shown in Table 2.4.

Table 2.3 refers to the publication of secondary documents that are not indexed in Scopus or WoS but are cited by indexed documents. The presence of Latin American authors from Argentina, Mexico, Brazil, Nicaragua, Honduras, and Bolivia stands out. It also reflects the commitment of researchers to reach indexed journals to make visible the epistemological approach that Latin America is gradually developing.

Table 2.4 shows that the journal Sustainability has been gaining prominence in the field of bioeconomy, followed by the Journal of Cleaner Production and New

Table 2.3 Top 10 LAC authors with the most articles and citations (2005–January 2024)

No	Author(s)	Secondary documents Scopus, WOS[a]	Dating
1	Trigo, E.	16	37
2	Rodríguez, A.	14	28
3	Enrique, G.	6	14
4	Bisang, R.	6	10
5	Mondaini, A.O.	5	17
6	Comberg, A.	3	14
7	Vera Morales, E.	3	6
8	Sol-Sánchez, A.	3	10
9	Zúñiga-González, C. A.	5	22
10	Catari-Yujra, C.	2	4

Source: Handmade
aSecondary documents not indexed in Scopus but cited by indexed documents

Biotechnology. This reflects those Latin American journals are not yet well positioned. However, the Ibero-American Journal of Bioeconomy and Climate Change of Nicaragua already appears as secondary documents.

Table 2.5 presents journals that are not indexed in Scopus or Web of Science, but whose papers are being cited by documents indexed in Scopus and WOS. This highlights the editorial efforts of Latin American journals to disseminate the epistemology of the bioeconomy. It is worth noting that the Ibero-American Journal of Bioeconomy and Climate Change stands out for its leadership, as it is cited in documents indexed in Scopus and WOS.

Based on the information processed and the review of the key authors and their research, the following considerations help to formalize the Latin American bioeconomy. Authors such as Bugge et al. (2016) and D'Amato et al. (2017) address all dimensions of sustainability, concluding that the bioeconomy is a benchmark for biotechnology, bioresources, and bioecology. However, like the pioneers, they omit other variables. While the European perspective is insightful, it should not exclude Latin American countries, which play an important role in the substitution process due to their raw materials. It is advisable to study the implications of introducing other variables that can be aligned with the proposals of the cited authors.

2.4 The Latin American Bioeconomy and Other Frontiers of Research

Quiroga and Zaiduni (2022) propose an articulating element between philosophical perspectives and applied public policies (involving thermodynamics) in a matter-energy relationship (where material aspects surpass energy aspects), establishing the basis for the convergence of ancestral and modern knowledge that Georgescu-Roegen omitted.

Table 2.4 Ten journals with the highest number of articles published in bioeconomy

Magazines	Wos	Scopus	Publications on the bioeconomy	Quartile	Country
Sustainability	79	83	106	Question 1	United Kingdom
Cleaner Production Magazine	44	47	61	Question 1	United Kingdom
New Biotechnology	47	41	75	Question 1	Netherlands
Biofuels, bioproducts and biorefinery	34	29	39	Question 1	United Kingdom
Biological resources technology	25	25	32	Question 1	United Kingdom
Forest policy and economy	24	24	30	Question 1	Netherlands
Economic Amphitheater	23	23	23	Question 1	Romania
Ambio	11	11	11	Question 1	Netherlands
Ibero-American Journal of Bioeconomy and Climate Change. UNAN-León	20	20	40	[a]	Nicaragua
Scandinavian Journal of Forest Research	17	10	21	Question 1	Sweden
Biomes bioenergy	8	8	9		United Kingdom
C3 Bioeconomy	2	2	4	[a]	Spain
ECLAC Review—Productive Development Series—Bioeconomy	0	0	0	[b]	Chile
Bioeconomy: A Bet on the Future of LAC	0	0	0	[b]	Colombia
Economy Short-term. Bioeconomy: A Review from Experience	1	1	2	[a]	Ecuador

Source: Authors, complemented by adaptation by Alzate Cárdenas et al. (2022)
[a]Secondary documents in Scopus
[b]Not indexed in Scopus or secondary documents

Thus, the Latin American Bioeconomy arises to complement the traditional bioeconomy, proposed by Georgescu-Roegen, establishing limits to economic exploitation to prevent the displacement of human beings in processes of high entropy that could be highly destructive (Scarlat et al., 2015). This approach is based on the Georgescu-Roegen mathematical model (Eq. 2.2) integrating the three variables mentioned: bioculture, bioterritory, and bioinformation.

Table 2.5 Ten journals with the highest number of articles published on the bioeconomy in Latin America

No	Magazines	Bioeconomy papers, publications, and secondary papers in Scopus and WOS[a]	Publishing institution	Country
1	Ibero-American Journal of Bioeconomy and Climate Change	20	National Autonomous University of Nicaragua, León.	Nicaragua
2	Argentine Bioeconomy Vision from Agribusiness	3	Ministry of Agroindustry	Argentina
3	Res. Soc. Dev	2	CDRR Editors	Brazil
4	The European Bioeconomy	2	Ministry of Agriculture, Fisheries and Food	Spain
5	Notebooks of the Spanish Society of Forest Sciences	2	Spanish Society of Forest Sciences	Spain
6	Ecoscience	1	EcoCiencia Foundation	Ecuador
7	Economy and Environment	1	World Bank	US
8	Short-term Economy	1	Gabriel René Moreno Autonomous University	Bolivia
9	Prospective Circular Economy and Research Bioeconomy	1	Global Web Icon University of Brescia	Italy
10	ECLAC	1	ECLAC	Chile

Source: Self-made
[a]Secondary documents in Scopus

$$Q_0(t) = F\begin{bmatrix} \frac{T}{R(t)} & \frac{T}{I(t)} & \frac{T}{M(t)} & \frac{T}{Q} & \frac{T}{W(t)} & \frac{T}{EF(t)} & \frac{T}{CA(t)} & \frac{T}{H(t)} & \frac{T}{K(t)} & \frac{T}{L(t)} & \frac{T}{TIC(t)} \\ 0 & 0 & 0 & 0 & 0 & 0 & 0 & 0 & 0 & 0 & 0 \end{bmatrix} \tag{2.3}$$

The new variables are as follows: *CA*, defined as bioculture or ancestral culture (which implies the convergence or combination of modern technology with traditional knowledge in a dialogue of knowledge that is summarized in bioculture); *EF*, defined as physical space (bioterritoriality); and *TIC*, referring to modern information and communication technologies derived from research and innovation, which according to the second law of thermodynamics, including *W* (waste introduced by Georgescu-Roegen), should not generate high entropy.

Explaining each of the variables, bioculture is a concept that encompasses cultural, social, and economic practices that promote sustainability and respect the environment and biodiversity. It focuses on promoting lifestyles in harmony with nature, encouraging organic farming, supporting the conservation of natural resources, and protecting cultural traditions related to the environment. This concept responds to the concerns of the green ecological scientists mentioned above, such as Scarlat et al. (2015). The concept is explained in detail by Quiroga and Zaiduni (2022).

Bioterritoriality allows us to analyze the role of the territory in local development processes (Dematteis & Governa, 2005, p. 32), where "territoriality becomes a main objective by seeking to clarify the connections between territoriality and sustainability, to define principles and models of territorial sustainability and to allow the sustainable autopoiesis of territorial systems."

This relationship between society and the environment, as well as the community dimension of the relations between the actors and their territory, according to Aceves González (1997), highlights the risk of a localist vision of development that must be analyzed multidimensionally to involve economic, social, cultural, political, and environmental aspects. Thus, local agents use and reproduce the territory, contributing to the construction of a new territoriality (Dematteis & Governa, 2005, p. 42).

Indigenous spiritualities also warn about the causes and effects of the processes of reconnection with nature that affect the political processes of self-determination of peoples. Therefore, indigenous peoples maintain cosmocentric worldviews and different epistemologies that deserve respect (Piedrahita Echandía et al., 2021, p. 140).

Finally, bioterritoriality is a "notion of territoriality applied to animal species, therefore, when it is related to the human species, it is inextricably linked to the notion of habitat" (Aceves González, 1997, p. 279).

Bioinformation is nourished by the information generated through bioinformatics. It is mainly used in medical sciences related to biotechnology; the production of renewable biomass; efficient bioprocesses; the production of biology, microbiology, biochemistry, and chemistry goods; as well as the production of goods and/or services related to bioinformatics (Gómez Restrepo, 2024). It is present in biomedical engineering, medical devices, and diagnostic reagents based on molecular biology, gene editing, microfluidics, genetic engineering, microbiology, biochemistry, biotechnology, and the production of goods and services in neurotechnology such as simulations of neural models, biological computers, brain-electronic systems interfaces, and activity measurement and analysis devices (Martínez, 2009; Santiago & Martínez 2003).

It should be remembered that the only multidimensional and complex model that simultaneously studies life or its sciences is bioeconomy (Zúñiga González, C. A., & Jaramillo-Villanueva 2022). It is complex and matrix like the one proposed by Georgescu-Roegen and even requires the use of information and communication technologies, where convergence involves the Internet of Things, cloud computing, blockchain, artificial intelligence, and many other elements that, if not controlled, can generate high entropy.

With the model in place, each major component of the equation is a flow or a reserve. All the information is multivariate and multidimensional and introduced matrix to complete the process of spatio-temporal restructuring that economic models failed to unify. In addition, it complements the matter-energy relationship due to the thermodynamic and low-entropy laws proposed by Georgescu-Roegen, whose equilibrium was known to ancestral peoples through an agricultural calendar called Chakana.

The epistemology of bioeconomy according to Zúniga González and Jaramillo-Villanueva (2022) stands out as a new interdisciplinary and holistic perspective that approaches production processes from an environmentally friendly perspective. Through a participatory approach in the university and rural community, an institutional discussion based on the interaction between the theories of biology, economics, and cognition begins. This approach recognizes the finite limits of natural resources and the capacity of the biosphere, relating them to the laws of thermodynamics and entropy.

The bioeconomic methodology uses data envelopment analysis (DEA) to measure productivity and technical efficiency, focusing on the availability of residual biomass. In addition, the bioeconomy contributes to understanding the coevolution of biological and socioeconomic systems, addressing social inequalities and the relationship between growth and entropy in the capitalist system, promoting a sustainable alternative approach (Stahel, 2010).

2.5 Conclusions and Recommendations

The Latin American bioeconomy is a paradigm that enriches Georgescu-Roegen's initial proposal while addressing the concerns of later scientists, whose problems were systematized with bibliometric methods.

Starting from a Western epistemological base, it introduces subjective precepts that enrich the proposal, making it innovative and potentially applicable initially in the Andes and the Amazon of South America, without limiting its replication in Central America and the Caribbean, where there are cultural features and epistemological aspects of the global South that are worth considering.

BL complements the traditional model of bioeconomy to adapt to the impacts of climate change by considering bioculture and bioinformation as fundamental axes, which do not align with variables that many analysts classify as part of Western epistemology and that are also the cause of the current capitalist crisis.

There is a potential risk that the economic agents involved in the bioeconomy will adopt utilitarian and consumerist precepts that distort the essence of the model as conceived by Georgescu-Roegen. This risk can be mitigated with flow variables such as the convergence of knowledge and territoriality that soften these entropic extremes.

The key to the proposal is to restore balance under new epistemological approaches where ancestral wisdom is valid and subjective, which, combined with modern objective technologies, can offer solutions that orthodox and heterodox currents have not yet found.

The need for Latin American governments and peoples to consider the enormous opportunity offered by the Latin bioeconomic approach is undeniable. It promotes sustainable development in harmony with nature, strengthens sovereignty within a democratic framework, and facilitates access to fundamental human rights underpinned by biocultural scientific knowledge for future generations.

Nicholas Georgescu-Roegen's research on the bioeconomy has led to the emergence of an emerging economic paradigm called "Latin American Bioeconomy" in Latin America and the Caribbean. This paradigm integrates Georgescu-Roegen's theory of energy value with three fundamental concepts: bioculture, bioterritoriality, and bioinformation. The results and discussion reveal that this new perspective not only enriches the European theory of energy value, but also responds to current societal needs. The epistemology of sentithought.

References

Aceves González, F. D. J. (1997). Territoriality nodal point at the intersection of urban space – communication processes – social movement. *Communication and Society – University of Guadalajara, 30*, 275–301.

Alzate Cárdenas, M. D. S., Guerrero Molina, M. I., & Gonzales Garcés, V. (2022). Bioeconomy. A systematic review and analysis from bibliometrics. *Revista En-Contexto, 10*(17). https://doi.org/10.53995/23463279.1246

Aksnes, D. W., & Sivertsen, G. (2019). A criteria-based assessment of Scopus and web of science coverage. *Journal of Data and Information Science, 4*(1). https://doi.org/10.2478/jdis-2019-0001

Bugge, M. M., Hansen, T., & Klitkou, A. 2016. What is the bioeconomy? A review of the literature. In Sustainability (Vol. 8, 7, p. 22). MDPI. https://doi.org/10.3390/su8070691

Carpintero, O. (2006). El desafío de la bioeconomía. Political Ecology, (30), 41–58.

Ceballos, H. Y. Z. (2019). Social re-enchantment with biocultural heritage through a design strategy for social innovation. *Encuentros de Investigación Formativa RAD, 2*, 196–207.

Ceballos Pérez, S. G., & Azamar Alonso, A. (2022). *Experiences and expectations of the bioeconomy*. UAM. https://repositorio.xoc.uam.mx/jspui/handle/123456789/47385

Chavarría, H. (2022). Strategies and policies for the bioeconomy in LAC, IICA. https://repositorio.iica.int/bitstream/handle/11324/22112/BVE24034060e.pdf?sequence=4&isAllowed=y

D'Amato, D., Droste, N., Allen, B., Kettunen, M., Lähtinen, K., Korhonen, J., Leskinen, P., Matthies, B. D., & Toppinen, A. (2017). Green, circular and bio economy: A comparative analysis of sustainability pathways. *Journal of Cleaner Production, 168*, 716–734. https://doi.org/10.1016/j.jclepro.2017.09.053

Dematteis, G., & Governa, F. (2005). Territory and territoriality in local development. The contribution of the slot model. *Boletín de La AGE, 39*, 31–53.

Echchakoui, S. (2020). Why and how to merge Scopus and web of sciences during bibliometric analysis: the case of the literature of the sales force from 1912 to 2019. *Journal of Marketing Analysis.* https://doi.org/10.1057/s41270-020-00081-9

Forero Lloreda, E., Rodríguez, C. A., & Rodríguez C., J. V. (2006). Transdisciplinary archaeology: A model of analysis in the management, conservation and dissemination of pre-Hispanic cultural and natural heritage in Colombia. *Bulletin of Anthropology University of Antioquia, 20*(37), 288–306. https://www.redalyc.org/articulo.oa?id=55703715

Georgescu-Roegen, B. (1966). *Analytical economics: Issues and problems* (1st ed.). Harvard University Press.

Georgescu-Roegen, N. (1967). Economic theory and agrarian economics. *The Economic Quarter, 34*(136), 589–638.
Georgescu-Roegen, B. (1971). *The law of entropy and the economic process.* Fundación Argentaria – Visor Distribuidores.
Georgescu-Roegen, N. (1974). Dynamic models and economic growth. En H. I. Greenfield, A. M. Levenson, W.B. Tollison, & S. F. Wachtel (Eds.), *The future of economic societies: Growth and regulation* (pp. 37–51). Praeger Publishers.
Georgescu-Roegen, N. (1975). Energy and economic myths. *Revista Económica del Sur, 41*(3), 347–381.
Georgescu-Roegen, N. (1977). What can thermodynamics and biology teach economists? *Atlantic Economic Journal, 5*, 13–21.
Georgescu-Roegen, N. (1979). Energy analysis and economic valuation (1979). In *From bioeconomy to degrowth: Georgescu-Roegen's "new economy" in eight essays* (pp. 4–44). Routledge.
Georgescu-Roegen, N. (1983). The energy theory of value: A particular economic sophistry. *El Trimestre Económico/Fondo de Cultura Económica, 50*(198), 829–860. https://www.jstor.org/stable/i23395067
Georgescu-Roegen, N. (1990). Production process and dynamic economy. In M. Beranzini & R. Scazzieri (Eds.), *The economic theory of structure and change* (pp. 198–224). Cambridge University Press.
Georgescu-Roegen, N. (1994). Nicholas Georgescu-Roegen on himself. In M. Szenberg (Ed.), *Great economists of today: The living testimony and worldview of today's great economists* (1st ed., pp. 149–186).
Gómez Restrepo, E. (2024). Design of a bioeconomic model based on the analysis of data on biodiversity and ecosystem services for the generation of sustainable development strategies in Colombia and the Andean Region [Doctoral thesis]. University of Antioquia, Medellín, Colombia.
Gould, H., Kelleher, L., & O'Neill, E. (2023). Trends and policies in the bioeconomy literature: A bibliometric review. *EFB Bioeconomy Journal, 3*, 100047. https://doi.org/10.1016/j.bioeco.2023.100047
Gudynas, E., & Acosta, A. (2011). The renewal of the critique of development and good living as an alternative. In International Journal of Ibero-American Philosophy and Social Theory (Vol. 53). http://www.cartalatinoamericana.com
Huang, T., Wu, H., Yang, S., Su, B., Tang, K., Quan, Z., & Luo, X. (2020). Global trends in research on sacral fracture surgery: a bibliometric study based on VOSviewer. *Thorn, 45*(12), E721–E728.
Hernández-Cervantes, T. (2008). Brief exposition of Georgescu Roegen's contributions to ecological economics and a critical commentary. *New Epoch, 56*, 35–52.
Martínez, M. (2009). Biocognitive theory: Towards a science of bioinformation. *Connections: Argentine Journal of Mental Health, 5*(17), 18–25.
Martínez-Alier, J. (2008). Conflits écologiques et langages de valorisation. *Écologie & Politique, 35*(1), 91–107. https://doi.org/10.3917/ecopo.035.0091
McDonough, W., & Braungart, M. (2003). Cradle to cradle: Redesigning the way we do things (A. García Brage & G. Pérez Van Kappel, Eds.; Spanish edition). McGraw-Hill/Interamericana de España S.A.U.
Meadows, D. H., Randers, J., & Meadows, D. L. (2012). The Limits to Growth: The 30-Year Update. Routledge.
Meadows, D., Meadows, D., Randers, J., & Behrens, W. (1972). The limits of growth: Report to the Club of Rome on the predicament of humanity (M. S. Loaeza De Graude, Ed.; 1st ed.). Fondo de Cultura Económica.
Mougenot, B., & Doussoulin, J. P. (2022). Conceptual evolution of the bioeconomy: A bibliometric analysis. *Environment, Development and Sustainability, 24*(1), 1031–1047. https://doi.org/10.1007/s10668-021-01481-2

Ordoñez Olivo, M. L., & Lakner, Z. (2023). Shaping the knowledge base of bioeconomy sectors development in Latin American and Caribbean countries: a bibliometric analysis. *Sustainability, 15*(6), 5158. https://doi.org/10.3390/su15065158

Perea Mosquera, L. N., Gaviria Arias, D., & Barrera Rodríguez, A. M. (2021). Bioeconomy: A bibliometric analysis in Latin America and the Caribbean. In *Risaralda competitive, generating research alliances for development* (pp. 1–27). https://doi.org/10.22517/9789587224955.2.1

Piedrahita Echandía, C. L., Perea Acevedo, A. J., & Useche Aldana, O. J. (2021). *Territorialities, Spiritualities and bodies: Critical perspectives in social studies* (1st ed.). Universidad Distrital Francisco José de Caldas – CLACSO: Editorial Magisterio.

Quiroga, J., & Zaiduni, M. (2021). From philosophy to public policy: Multidimensional index for living well and the bioeconomy (1st ed.). Editorial Académica Española.

Quiroga, J., & Zaiduni, M. (2022). The economy of life: The Amazonian Andean bioeconomy (D. Valdivia Coria, Ed.; 1st ed.). Industrias Gráficas Inventados SRL.

Quiroga-Canaviri, J. L., & Menéndez-Gámiz, C. R. (2023). From Georgescu-Roegen's bioeconomy to the Andean Amazonian bioeconomy. *C3-Bioeconomy Circular and Sustainable Bioeconomy, 4*, 25–54. https://doi.org/10.21071/c3b.vi4

Ramírez-Montoya, M., & García-Peñalvo, F. (2018). Co-creation and open innovation: Systematic literature review. [Co-creation and open innovation: Systematic literature review]. *Comunicar, 54*, 09–18. https://doi.org/10.3916/C54-2018-01

Santiago, J. L., & Martínez, M. E. (2003). Biocognición de los receptores de las membranas celulares: de la biocultura a la citocultura. *Psiquiatria.com, 7*(3).

Scarlat, N., Dallemand, J. F., Monforti-Ferrario, F., & Nita, V. (2015). The role of biomass and bioenergy in a bioeconomy of the future: Policies and facts. *Environmental Development, 15*, 3–34. https://doi.org/10.1016/j.envdev.2015.03.006

Sen, A. (2014). Development as freedom (1999). In *The globalization and development reader: Perspectives on development and global change* (p. 525). John Wiley & Sons.

Stahel, W. R. (2010). The economics of performance (2nd ed.). Palgrave Macmillan.

Van Eck, N., & Waltman, L. (2010). Software survey: VOSviewer, a computer program for bibliometric mapping. *Scientometrics, 84*(2), 523–538.

Zhu, J., & Liu, W. (2020). A history of two databases: The use of web of sciences and Scopus in academic articles. *Scientometrics, 123*(1), 321–335. https://doi.org/10.1007/s11192-02003387-8

Zúñiga-González, C. A. (2023). Impact of the post-Covid-19 TFP bioeconomy on the agricultural economy. *PLoS One, 18*(11), e0288885. https://doi.org/10.1371/journal.pone.0288885

Zúniga-González, C. A., López, M. R., Icabalceta, J. L., Vivas-Viachica, E. A., & Blanco-Orozco, N. V. (2022). Epistemology of bioeconomics. *Ibero-American Journal of Bioeconomy and Climate Change, 8*(15), 1786–1796. https://doi.org/10.5377/ribcc.v8i15.13986

Zúñiga González, C. A., & Jaramillo-Villanueva, J. L. (2022). Frontier model of the effects of environmental inefficiency on the livestock bioeconomy [version 1; peer review: awaiting peer review]. *F1000Research (Talk) 2022, 11*, 1382. https://doi.org/10.12688/f1000research.128071.1

Zuniga-Gonzalez, C. A., Quiroga-Canaviri, J. L., Brambila-Paz, J. J., Ceballos-Pérez, S. G., & Rojas-Rojas, M. M. (2024). Formulation of an innovative model for the bioeconomy. *PLoS ONE, 19*(11), e0309358. https://doi.org/10.1371/journal.pone.0309358

Chapter 3
Life Cycle Analysis as a Tool for the Evaluation of the Environmental Impact of a Coffee Residual Mixture Biorefinery

Alejandro Téllez Jurado, Eva Luz Hernández-Teyssier, Sergio Gabriel Ceballos Pérez, and Miguel Angel Anducho-Reyes

Abstract Life cycle analysis (LCA) (using the Ecoinvent 3.2 database, the ReCiPe 2016 Midpoint H method, and OpenLCA 2.0 software) can be used as a possible tool to determine the environmental impact of a coffee residual mixture biorefinery. Five scenarios were defined, which represented the products obtained from the biorefinery; in scenario (Sc) 1, the products were polyphenols and lignin, Sc2 a mixture of sugars, Sc3 a crude enzymatic extract, Sc4 another crude enzymatic extract, and Sc5 a mixture of sugars. It was observed that Sc4, defined as the integrated scenario since it was formed from stages from other scenarios, had the lowest environmental impact. At the same time, Sc1, 2, and 5 had the highest environmental impact due to the use of energy, synthesis of reagents, and chemical residual generated. It was observed that the LCA provides valuable information for decision-making on the environmental impact of processes involving the valorization of lignocellulosic residual, such as the mixing of coffee residual study.

Keywords Circular economy · Coffee residual mix · Environmental impact · Life cycle analysis · Valorization

A. Téllez Jurado (✉) · E. L. Hernández-Teyssier · M. A. Anducho-Reyes
Laboratorio de Aprovechamiento Integral de Recursos Bióticos, Universidad Politécnica de Pachuca, Zempoala, Mexico
e-mail: alito@upp.edu.mx; eteyssier@upp.edu.mx; anducho@upp.edu.mx

S. G. Ceballos Pérez
Dirección de Investigación, Innovación y Posgrado, SECIHTI-Universidad Politécnica de Pachuca, Zempoala, Mexico
e-mail: ceballosp@upp.edu.mx

S. G. Ceballos Pérez et al. (eds.), *Sustainability in Latin America*,
https://doi.org/10.1007/978-3-032-09787-3_3

3.1 Introduction

Agribusiness can promote and achieve global economic, social, and environmental development if the activities developed and the protection of the environment are maintained in balance through its processes, from the handling of raw materials to the distribution and final disposal of by-products (Bichyova et al., 2019). The ineffective disposal of these agro-industrial residuals represents high costs, generating not only an environmental problem but an economic one due to the lack of profits from the investment made to reduce pollution (Yepes et al., 2008). Using agro-industrial residuals through innovative technologies allows for solving different environmental problems originating from the generation and disposal of these residuals and other economic factors (Cury et al., 2017; Bichyova et al., 2019; Zuniga-González, 2024).

In recent years, at a national level, the solid residual generated has exceeded 10,000 tons per day, and more than 74% corresponds to organic residual, of which only 3.6% is treated to value or reduce its volume of contamination (INEGI, 2014). The technologies used to exploit this have been integrated into the same biorefinery concept. This allows them to be classified as processes through which industrial and environmental interest products can be obtained while reducing the contaminating organic matter.

Coffee processing generates large amounts of waste that, by its nature, is considered recalcitrant. This waste is not used due to toxicity, so it is not treated correctly and is confined in the open air or underground pits. Coffee waste has a high content of various molecules, with holocellulose and phenolic compounds standing out. These characteristics make coffee waste a potential source of multiple compounds that can have a high added value. The coffee biorefinery arises as an opportunity to use the waste that, when integrated into the coffee processing, would favor the environment and the productive, social, and economic sectors. The study of the life cycle of a coffee biorefinery can provide information to generate an integrated process through the concept of bioeconomy and minimize the impact of coffee processing.

3.2 Coffee Waste and Its Biorefinery

3.2.1 The Biorefinery

Currently, there is no precise and concrete definition of the biorefinery; however, some authors have mentioned its relationship and analogy to the oil refinery. For this reason, we propose a definition derived from the previous concepts:

> Any activity integrated into the processing of one or several biomass sources to obtain materials or products with the primary objective of fully utilizing the organic matter with the least possible environmental impact.

The degree of integration that the biorefinery can achieve will allow the same degree of utilization of the residual or biomass. For this purpose, three integration categories have been defined: first-, second-, and third-generation biorefinery (Kamm et al., 2008; Gomez-Millan, 2015).

The first-generation biorefinery uses biomass sources such as sugar cane, corn, soybeans, etc., from which fermentable sugar solutions, mainly bioethanol or oils, are easily obtained for biodiesel production. These are the least sophisticated in terms of utilization since they require raw materials concentrated in a substance and do not pose threats such as crop consumption (Kamm et al., 2008).

The second-generation biorefinery focuses on using lignocellulose as a raw material and its three main components: cellulose, hemicellulose, and lignin. The products to be obtained are varied, but the main problem is the fractionation of its parts. The main advantage is their abundance since they represent renewable carbon sources and do not compete with any primary need, such as food; however, this type of residual in the biorefinery improves the use and disposal of energy and the recycling of residual (Kamm et al., 2008; Carrillo-González & Ponce-Sánchez, 2019).

Integrated or third-generation biorefineries would produce, through multiple technologies, product lines (biofuels, chemicals, plastics, etc.) from biomass of diverse origin. They are the most ambitious because they extend to a greater diversity of residuals from agriculture, livestock, urban solid residuals, petrochemicals, paper mills, etc., consequently expanding the range of products, such as ethanol, fuels, agrochemicals, and plastics. The advantages of this biorefinery are the possibility of using almost any organic residual, the lack of competition with previous forms of biorefinery, and the production of a clean form of energy. Forecasts agree that, with them, it will be possible to take advantage of all the possibilities offered by biomass, reducing the generation of residual and making the use of this residual more sustainable (Kamm et al., 2008; Cherubini et al., 2009; Carrillo-González & Ponce-Sánchez, 2019).

At a global level, two coffee species are the most exploited: *Coffea arabica* (Arabica) and *Coffea canephora* (Robusta) (International Coffee Organization, 2020). The first species represents 75% of the total production due to its cultivation conditions (from 600 to 2000 masl), and the second species provides the remaining 25% of the world production since it can only be cultivated at less than 600 masl; however, it is more resistant to pests (Mussatto et al., 2011; Sánchez & Anzola, 2012).

3.2.2 Coffee Processing

The main objective of processing the coffee cherry is to remove the pulp, mucilage, parchment, and skin to expose the bean. After the cherry is harvested, coffee is commonly processed by two different methods: wet and dry processing (Oliveira & Franca, 2015).

In the dry process, also called the dirty or natural coffee process, the fruit is exposed to the sun until it reaches a humidity of 10–11%, which takes about 15–20 days of air drying. The dried cherries obtained are called coffee balls. Initially, the grain is dispersed on the ground, concrete, or asphalt during the day and collected at night. This process is repeated every day until dry. Later, through mechanical processing, the outer layers of the fruit are removed, and a by-product called coffee bagasse is generated (Pandey et al., 2000; Murthy & Madhava, 2012; Hikichi et al., 2017). The green coffee bean is distinguished by its sweet and soft aroma, acquired by impregnating sugars or compounds in the mucilage during drying. This method is commonly used in the Robusta species and involves the fermentation of the bean (Aristizábal-Marulanda et al., 2017; Janissen & Huynh, 2018).

During the wet process, mainly applied to the Arabica species, the skin and pulp of the cherry are mechanically removed until the coffee bean has a viscous texture, similar to honey, which will disappear when the bean is fermented for 24–48 h; the time will depend on the climatic conditions, coffee ripening, amount of mucilage, and volume of processed coffee (Aristizábal-Marulanda et al., 2017). After fermentation, the beans are washed for 12 h in clean water. The final bean acquires a better quality due to removing terpenes and polyphenols. Finally, the bean is dried until it reaches a moisture content like that of the dry method (Murthy & Madhava, 2012), a product known as parchment coffee because the green coffee beans are covered by their parchment (endocarp) and silver skin (epidermis) (Aristizábal-Marulanda et al., 2017; Hikichi et al., 2017). The method selection depends on the site conditions to best process the cherry; for example, in regions with high relative humidity, prolonged drying would not work as well as machine pulping (Aristizábal-Marulanda et al., 2017).

Due to the high water consumption in the wet method, processing alternatives have been proposed, through which less than 1 l of water could be required per kilogram of parchment coffee produced (Mussatto et al., 2011). Finally, as an innovation for the reduction of contaminants, a third process known as semi-dry or semi-humid can be carried out, through which, by mechanical effect and large quantities of water, the layers of the fruit are removed, generating a large volume of residual from bagasse, pulp, and residual water (Hikichi et al., 2017).

3.2.3 Coffee Wastes

According to the International Coffee Organization (2020), the coffee industry generates a considerable amount of coffee products worldwide. Because of its industrial processing, a large volume of residual or by-products is generated in the various processing stages; the said residue is considered highly polluting and is rarely disposed of responsibly.

Coffee husk (CH) is a skinny outer layer attached to the green coffee bean, generated in the roasting process, and represents about 4.2% of the total weight of the fruit (Farah & Dos Santos, 2015; Janissen & Huynh, 2018). The structural composition of SC, according to Ballesteros et al. (2014), on a dry basis, is 23.77 ± 0.09% cellulose, 16.68 ± 1.30% hemicellulose, 28.58 ± 0.46% total lignin, in which 20.97 ± 0.43% is insoluble and 7.61 ± 0.16% is soluble; in terms of sugars, it has 58.76% glucose, 18.81% xylose, 9.29% galactose, 8.75% arabinose, and 4.37% mannose. In addition, the mineral content was determined, the highest concentrations (mg/kg dry weight) being potassium (21.1 ± 0.00), calcium (9.4 ± 0.01), magnesium (3.1 ± 0.00), sulfur (2.8 ± 0.00), phosphorus (1.2 ± 0.00), iron (843.30 ± 7.90), and aluminum (470.60 ± 13.9). These values may vary depending on the species due to growing conditions, type of sowing, and soil characteristics.

When the cherry is processed by the wet method, the residue obtained is coffee pulp (CP), and when it is processed by the dry method, coffee bagasse (CB). Both cases are the same layer of the cherry; however, the CP represents 41% of the residue and the CB 18%; so, by volume of residual generated, the wet process generates more contamination (Murthy & Madhava, 2012). Both the CP and the CB are rich in carbohydrates (44–57.8% w/w), contain 43–63% w/w of cellulose, 2.3–7% of hemicellulose, and a large amount of tannins (1.8–8.6% in pulp and 4.5–9.3% in bagasse) and a considerable amount of caffeine (1.2–1.3% w/w), which is why these residues have high toxicity and low biodegradability and are not very useful as livestock feed. However, it could be used as a substrate for producing edible mushrooms (Pandey et al., 2000; Murthy & Madhava, 2012; Janissen & Huynh, 2018).

3.3 Methodology

3.3.1 Evaluation of the Biorefinery Through Life Cycle Analysis

Life cycle analysis (LCA), also known as cradle-to-grave analysis, is a methodology developed from a request from the Coca-Cola company to Harry Teasley in 1969 to find less polluting beverage containers, known as Resource Analysis and Environmental Profile (RAEP) (Gnansounou & Pandey, 2017), to be later considered for process standardization in the 1990s, due to the goal of Sustainable Development; therefore international organizations including the International Organization for Standardization (ISO) captured it in a series of standards, ISO 14040 series (ISO, 2006), as a guide to assess the environmental impacts of a product through the life cycle including raw material, production, use, treatment, recycling, and final disposal, considering it a holistic methodology (Gnansounou & Pandey, 2017).

According to ISO 14040, LCA follows four stages: (1) definition of the objective and scope, (2) inventory analysis, (3) impact analysis, and (4) interpretation (ISO, 2006). The first stage is critical because it clearly defines the aim of the study and the scope, the target people, and the intended uses of the results since the interpretation will be based on these. Therefore, defining the functional unit is essential to develop the next stage. The functional unit is the basis for calculating the inventory and impact analyses; this can be a defined quantity of final product or raw material, depending on the focus of the study (ISO, 2006; Gnansounou & Pandey, 2017). To develop the second stage, inventory analysis, the system units are first described as the individual processes of the complete process and are detailed as much as required by the LCA. To do this, the inputs and outputs of each unit are declared, considering the material and energy balances significant and less critical (Gnansounou & Pandey, 2017). The information on inputs and outputs can be acquired in the process or estimated by similar processes but validated to simulate the impact of the study system as closely as possible (Schmidt-Rivera et al., 2020).

In stage three, the input or output substances and the equipment used for the system units are studied from the perspective of environmental impact; therefore, it is necessary to take care of the impact model to be applied since this stage is considered essential in the LCA (ISO, 2006), because the derived results are focused according to the scope and objective of the LCA (stage one). Therefore, two impact categories are differentiated, the end-point category and mid-point category (Gnansounou & Pandey, 2017); currently, both categories are found as impact methods in free software (Green Delta, 2023) and licensed software (CADIS, 2011), and they include all the categories accepted by the ISO series and other international entities. Finally, in stage four, the results are interpreted according to the objective and scope of the LCA (ISO, 2006), taking into account the consistency, integrity, and sensitivity to relevant uncertain factors since the results can be interpreted from the limits of the system, the functional unit, the temporal or spatial scope, and the sensitivity to the reference system, giving high or low weight to the impact categories of the selected impact methods or in compliance with a regulation (Gnansounou & Pandey, 2017).

This is why the LCA is a necessary tool for evaluating the design, application, and evolution of biorefineries (Schmidt-Rivera et al., 2020), from the design of cutting-edge processes to their environmental performance and primary, pilot, or demonstrative research (Carrillo-González & Ponce-Sánchez, 2019). It also systematically builds the necessary knowledge to discriminate between the appropriate biorefinery options and contribute to generating energy and environmental policies (Caldeira et al., 2020).

3.4 Results

3.4.1 Valorization of Agro-industrial Residual by LCA

Global initiatives focused on creating economic valorization chains that encourage the efficient use of renewable resources and the reduction of pollutants have promoted the valorization of agro-industrial residual (Carrillo-González & Ponce-Sánchez, 2019; Caldeira et al., 2020). Therefore, the economic model must change from linear to circular. For this, the biorefinery would be the means and life cycle analysis tool for designing the processes and products to be integrated into the new economy (ONU, 2018; López-Arenas & Anaya-Reza, 2019).

The circular economy emerged conceptually from various social demands calling for the care and respect of the environment (Carrillo-González & Ponce-Sánchez, 2019; Zuniga-González & Jaramillo-Villanueva, 2022). So, in 1980, the then Minister of Norway, H. Brundtland, presented a document, "Our Common Future," at the United Nations World Commission on Environment and Development, in which she planted in humanity the idea of a possible harmony between economic growth and the use of natural resources, giving rise to the concept of Sustainable Development (Carrillo-González & Ponce-Sánchez, 2019; López-Arenas & Anaya-Reza, 2019).

However, it was not until 2006 that the Organization for Economic Cooperation and Development (OECD) began to disseminate the sustainable development approach through the document "The Bioeconomy to 2030. Designing a Policy Agenda," for which they propose the creation of new technologies and new professions in response to the future demands of a circular economy or sustainable development, such as biorefinery and biotechnology (OECD, 2006). The approach of the circular economy, according to this document, is to generate manufacturing processes with zero residual when biomass is used as raw material, considering biorefinery as necessary for the generation of everyday products such as biofuels, electricity, bioplastics, etc., reducing the environmental impact of processes and products by replacing renewable raw material and highly demanded value-added products (Rajesh Banu et al., 2020; Solarte-Toro & Cardona Alzate, 2021). For biorefineries to fulfill their objective of being a means towards sustainable development and care for the environment in a circular economy (Aristizábal-Marulanda et al., 2017), three aspects must be considered: (1) availability of biomass over time, (2) analysis of technical-economic potential, and (3) environmental assessment based on the life cycle (Caldeira et al., 2020). Several studies have been developed for the use of agro-industrial residual in terms of biorefinery (Battista et al., 2020; Solarte-Toro & Cardona Alzate, 2021); however, aspects 2 and 3, mentioned above, still need to be developed in the biorefinery to consider it as a means to reach the circular economy, since in principle biotechnological processes are considered eco-friendly; however, certain raw materials or energy demands detract from the value of the generated product (Schmidt-Rivera et al., 2020). Therefore, for the systematization and implementation of biorefineries, it is suggested to complete the studies with the following aspects: (2) analysis of technical-economic potential and (3) environmental assessment based on the life cycle (Caldeira et al., 2020).

3.4.2 *Valorization of a Coffee Residual Mixture Biorefinery by LCA: Case Study*

The life cycle analysis (LCA) must be carried out according to the phases described by ISO 14040: Environmental management—Life cycle analysis, which is: Definition of objectives and scope of the study system, Inventory analysis, Impact assessment and Interpretation (ISO, 2006).

The LCA can focus on evaluating and comparing the environmental impact associated with using agricultural residual using the biorefinery concept, in this case, of a coffee residual mixture (CRM). It is essential as a first step to determine the scope of the biorefinery, that is, the products that will be obtained from the various processes to which the coffee residual will be subjected (Fig. 3.1). Each product will generate a scenario that will be evaluated based on the existing inventory: fuel, energy, reagent, transportation, etc. Finally, it is also necessary to define the Fundamental Unit (FU), which indicates the mass with which the process will begin to achieve a mass balance of the total process at the end. The information from each scenario will allow the environmental impact of each of them and the total process to be defined. This study considered five scenarios, starting from 100 kg as

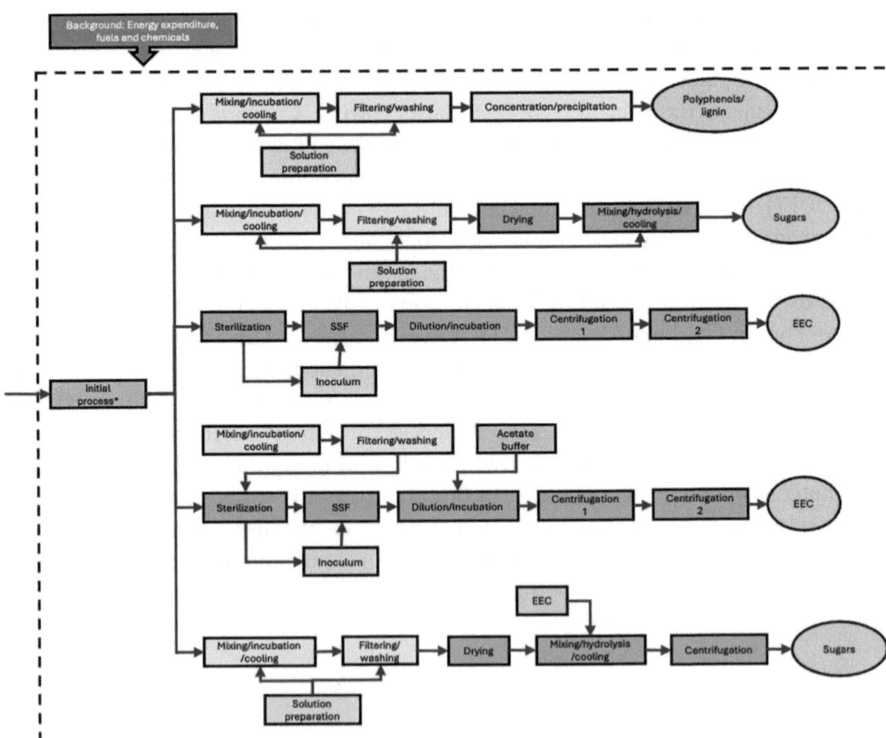

Fig. 3.1 Breakdown scenarios for coffee waste biorefinery

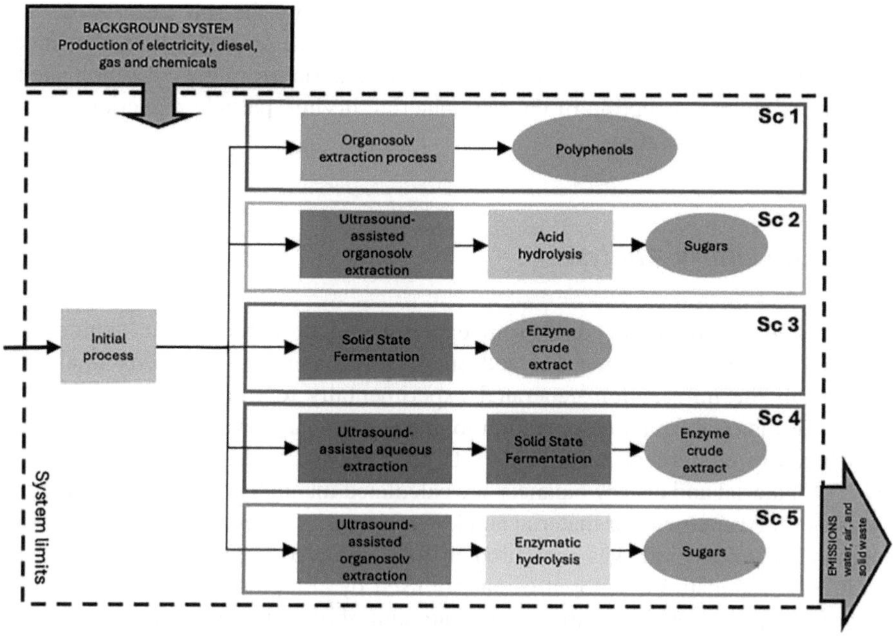

Fig. 3.2 Identified scenarios for the coffee waste biorefinery. Each block color represents different operations. Blocks with the same color are the operations that are repeated in each scenario

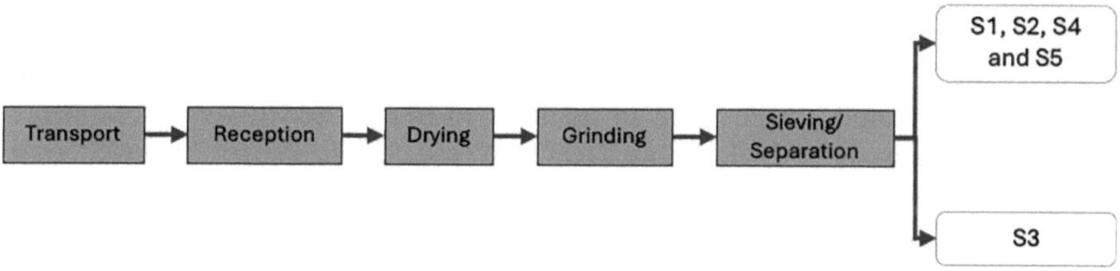

Fig. 3.3 Operations involved in the initial process

the FU of CRM. Each scenario represented a by-product of the process. The system is limited to the collection and transportation of the CRM until the products are in crude solution, without purification. The products obtained can be purified to obtain finer products with greater added value. Hence, it is essential to define the scope of the biorefinery to determine with greater precision the possible impact of each scenario or the total process.

Figure 3.2 shows the proposed CRM biorefinery system. The five scenarios studied can be seen. Each scenario generates a crude product, representing the final product for this study. Thus, Sc 1 generates polyphenols and lignin; Sc 2 generates a mixture of unpurified sugars; Scenario 3 generates a Crude Enzyme Extract (CEE); Sc 4 also generates a Crude Enzyme Extract; and finally, Sc 5 generates a mixture of unpurified sugars as the final product. Fig. 3.3 shows the initial process that includes the previous activities for preparing the raw material, in this case, the CRM. Fig. 3.3 shows the operations that are carried out in the initial process. This stage includes transporting the raw material from the collection site to the processing site, reception of the raw material, drying, particle reduction (grinding), sifting, and separation.

3.4.3 Inventory Analysis

In this stage of the LCA, data from experimental or bibliographic sources are collected to perform the material and energy use balances. Therefore, this stage requires all the information generated experimentally to calculate the yields and transfer of material from one operation to another and calculate the energy spent by the laboratory equipment according to its capacity and material considered in the FU. The material and energy balances are calculated manually, considering the kg as the unit of measurement for material and kWh for energy. Specifically, in the transfer operation, the transport was measured in tons transported per kilometer traveled (t*km). To calculate the amount of energy used by equipment, the work capacity, time of use, power of the equipment, and amount of material treated are considered, and according to the scope of this system, energy is not recovered, and the primary source of energy is the national electrical substation.

In cases where the production of a resource used in the scenarios is not an operation carried out experimentally in the laboratory or by the work team, the Ecoinvent 3.2 database (2015) is used to perform cut-off type assignments. These resources were electric power, diesel, gas, and chemicals (ethanol, sulfuric acid, deionized water, and acetic acid), in addition to being used to assign elemental output flows such as PE, RLq, and RRSO (Wernet et al., 2016); however, even though it is considered a complete database, it is necessary to add a stage for the preparation of buffer solutions, as well as an operation for the chemical synthesis of the components of the buffer that will be used (for the generation of EEC). Table 3.1 lists the names of the additional flows taken from the database, which were not measured in the laboratory, accompanied by their ID. With the information obtained, an input-output balance is made for each operation and each scenario.

3.4.4 Environmental Impact Assessment Analysis

The life cycle environmental impact analysis (LIACV) is developed using the ReCiPe 2016 Midpoint (H) method with the OpenLCA 2.0 software (Green Delta, 2023) considering the response of the 18 categories grouped into damage to human health: stratospheric ozone depletion (SOD), fine particle formation (FPMF), ozone formation-human health (OF-HH), carcinogenic human toxicity (HCT), non-carcinogenic human toxicity (HnCT), and ionizing radiation (IR); global

Table 3.1 Additional flows obtained from the Ecoinvent 3.2 database

Stream name	Ecoinvent 3.2 ID
Diesel production, low-sulfide-RoW	b80eb2d5-1256-41bb-a61d-83252d0dfd4
Transport, tractor, and trailer, agriculture-RoW	ba4b7781-39f2-465b-8d98-0bc1c4c82e6
Electricity, high voltage, mixed production-MX	66c93e71-f32b-4591-901c-55395db5c13
Ethylene hydration-RoW	3a0d77a5-6488-4977-966a-731350be864
Sulfuric acid production-RoW	9d63da75-8289-4b96-a900-67ec3bd40a1
Production of deionized water from tap water for the user-RoW	360e2eb0-f81c-4e4b-ba6b-c7a690f3127
Natural gas production-RoW	28a32250-3ba9-4556-9505-8841cff28aa
Butane oxidation-RoW	e36aa595-9fc6-4eff-8f34-a9b11f46076

warming: global warming (GW) and water consumption (WC); damage to ecosystems: freshwater ecotoxicity (FE), marine ecotoxicity (ME), terrestrial ecotoxicity (TE), freshwater eutrophication (FEu), marine eutrophication (MEu), terrestrial acidification (TA), ozone formation-terrestrial ecosystems (OF-TE), and land use (LU); resource availability: fossil resource scarcity (FRS) and mineral resource scarcity (MRS) (Huijbregts et al., 2016) and the hierarchical value system (H) because it is the most recommended by the International 100-year Reference Life Cycle Data System (Gnansounou & Pandey, 2017). The information provided by the OpenLCA software is processed and statistically analyzed.

3.4.5 *Environmental Impact Analysis of the Process by ReCiPe 2016*

By entering all the values of the flows into the specialized software, the values of each category of the ReCiPe method are obtained for each scenario tested. Initially, the ReCiPe 2016 method had 17 midpoint impact categories; however, the software returns 18 since the freshwater eutrophication potential divides it into marine eutrophication (MEu) and freshwater eutrophication (FEu). Therefore, the 18 midpoint categories are shown. Considering the interpretation of the impact through the categories to be necessary, they were grouped into damage to human health, climate change, damage to ecosystems, and damage to the availability of resources since the general interpretation of each category is linked to these aspects, which can be seen from the names of each category shown in Table 3.2. For example, the categories of the damage to human health group show values of indicators directly related to the

Table 3.2 Results of the life cycle environmental impact analysis

	Reference unit	Sc1	Sc2	Sc3	Sc4	Sc5
Damage to human health						
Stratospheric ozone depletion (SOD)	kg CFC11 eq	4.2E-05	5.3E-05	8.0E-05	7.2E-05	13.1E-05
Formation of fine particles (FPMF)	kg PM2.5 eq	2.6E-01	3.6E-01	3.6E-01	3.2E-01	6.6E-01
Ozone formation-human health (OF-HH)	kg NOx eq	13.3E-01	8.55E-01	4.14E-01	3.73E-01	10.8E-01
Human carcinogenic toxicity (HCT)	kg 1,4-DCB	3.3E+00	4.6E+00	2.3E+00	2.1E+00	5.8E+00
No-human carcinogenic toxicity (HnCT)	kg 1,4-DCB	5.5E+01	7.8E+01	3.7E+01	3.3E+01	9.6E+01
Ionizing radiation (IR)	kBq Co-60 eq	0.67E+01	0.85E+01	1.4E+01	1.3E+01	2.2E+01
Climate change						
Global warming (GW)	kg CO_2 eq	2.0E+02	2.8E+02	1.9E+02	1.6E+02	4.0E+02
Water consumption (WC)	m^3	0.81E+03	0.84E+03	2.0E+03	1.8E+03	3.1E+03
Damage to ecosystems						
Fresh water ecotoxicity (FE)	kg 1,4-DCB	2.1E+00	3.1E+00	1.2E+00	1.1E+00	3.5E+00
Marine ecotoxicity (ME)	kg 1,4-DCB	3.0E+00	4.3E+00	1.9E+00	1.7E+00	5.1E+00
Terrestrial ecotoxicity (TE)	kg 1,4-DCB	2.2E+02	2.9E+02	2.7E+02	2.4E+02	5.2E+02
Fresh water eutrophication (FEu)	kg P eq	6.1E-02	8.8E-02	3.1E-02	2.8E-02	9.5E-02
Marine eutrophication (MEu)	kg N eq	1.3E-02	1.7E-02	2.2E-02	2.0E-02	3.7E-02
Terrestrial acidification (TA)	kg SO_2 eq	0.78E+00	1.1E+00	1.1E+00	0.97E+00	2.0E+00
Ozone formation-terrestrial ecosystems (OF-TE)	kg NOx eq	19.3E-01	10.7E-01	4.2E-01	3.8E-01	12.4E-01
Land use (LU)	m^2 annual crop eq	2.1E-01	3.0E-01	1.7E-01	1.6E-01	4.0E-01
Damage to resource availability						
Fossil resource scarcity (FRS)	kg oil eq	12.8E+01	18.4E+01	6.8E+01	6.2E+01	20.4E+01
Mineral resource scarcity (MRS)	kg Cu eq	3.7E-01	5.4E-01	1.3E-01	1.1E-01	5.1E-01

reduction of ozone or environmental quality that affect human health in various ways. The values of each category are presented based on the FU (100 kg wet weight of the collected residual). In the categories of the damage to human health group, the smallest values of all the categories are found, which are for the stratospheric ozone depletion (SOD) category in the range of 4.17 × 10–5 to 1.31 × 10–4 kg CFC11 eq. In addition, in this group, in five of the six categories, Sc 5 presented the highest values, leaving Sc 1 (physicochemical processing) with the highest value obtained in the remaining category, ozone formation-human health (OF-HH) with 1333 kg NOx eq.

The categories of the climate change group are those with the highest values and, in some cases, the categories with the greatest attention in the LCAs. On the one hand, the highest value of global warming (GW) was again shown by Sc 5 with a total of 398,953 kg CO_2 eq, followed by Scenario 2 (Integrated processing) with 278,989 kg CO_2 eq. In the other category, water consumption (WC), the values range from 810,915 m^3 to 3,053,942 m^3, with Sc 5 again having the highest value and, in this case, Sc1 having the lowest impact for this category.

There are eight categories in the damage to ecosystems group, which are mainly associated with the deterioration of terrestrial and aquatic ecosystems and soils; among which, terrestrial ecotoxicity (TE) stands out because it presented the highest values, in the range of 215,125 to 515,581 kg 1,4-DCB, with Sc 5 again having the highest value and Sc1 the lowest. Continuing with the categories focused on ecotoxicity, this group also includes freshwater ecotoxicity (FE) and marine ecotoxicity (ME), which show a maximum value of 3507 kg 1,4-DCB and 5115 kg 1,4-DCB, respectively, for Sc 5. From the perspective of the eutrophication study, this group includes the two categories mentioned at the beginning of this section, FEu and MEu; both categories presented small values compared to other categories in this group; for example, in FEu, Sc 4 (integrative processing) showed the minimum value of the category with 0.028 kg P eq, followed by Sc 3 (Biological processing) with 0.031 kg P eq; on the other hand, in MEu, Sc 1 obtained the minimum value of 0.013 kg N eq and Sc 5 the maximum of the category by obtaining 0.037 kg N eq. Continuing with the other categories in this group, in the categories terrestrial acidification (TA) and land use (LU), Sc 5 obtained the highest value of 1976 kg SO_2 eq and 0.402 m^2 annual crop eq, respectively, while for ozone formation-terrestrial ecosystems (OF-TE) Sc 1 had the highest value, presenting 1932 kg NOx eq.

Finally, the last grouping, damage to resource availability, comprises fossil resource scarcity (FRS) and mineral resource scarcity (MRS). In these categories, Sc 4 showed the lowest values with 61.738 kg oil eq for FRS and 0.114 kg Cu eq for MRS, and the highest values for FRS were achieved by Sc 5 (204.258 kg oil eq) and for MRS by Sc 2 (0.537 kg Cu eq).

This type of work presents an excellent opportunity to enrich knowledge in the area of study of the use of coffee residues evaluated by LCA, through which the diversity in the type of UF, number of scenarios, and processes can be observed. This information helps in choosing the impact method for the LCA and the characteristics of the Ecoinvent inventories and flows to direct the LCA toward greater homogeneity. On the other hand, the chosen base scenarios were considered acceptable due to the strategies developed for using residual and its evaluation by LCA. However, comparing the studies is complex and complicated due to the need for more homogeneity in FU and/or product of interest. For example, making the comparison of Sc 3 and Sc 4 with the results obtained by Catalán et al. (2019) is not objective since they use a smaller FU, and these authors evaluated based on the purified final product (enzyme) and the use of by-product (compost). However, it can be mentioned that the values acquired are, in all categories, much lower.

Likewise, the comparison with the study carried out by Schmidt-Rivera et al. (2020) is also not possible due to the magnitude of the FU, which is one ton of residual, in addition to a great diversity of processes used and products, such as the generation of biodiesel, processes with energy recovery, compost generation, and direct use in soils, which are very different from the products of the study. In this sense, the utilization approaches are considered as the point of comparison between the studies, which is why the five scenarios of the present study show lower values in nine categories, which include while in SOD, FPMF, HnCT, GW, TE, MEu, TA, FRA, and MRS; on the other hand, in OF-HH, HCT, FE, ME, Feu, and OF-TE, it has values very similar to those described by Schmidt-Rivera et al. (2020). Therefore, sometimes a utilization method reduces environmental impact while, on the other hand, that same method increases other impacts.

3.4.6 Study Prospects

The first prospect, if the use is through one of the first two scenarios, will be to use an extraction procedure with lower ethanol or sulfuric acid reagent requirements or the use of ethanol produced through a less polluting process since the mixing/incubation/cooling operation had an effect in 12 and 13 categories, in Sc 1 and Sc 2, respectively. This coincides with that reported by De Marco et al. (2018), whose supercritical CO_2 extraction process affects the FE, ME, and LU categories, whose values are 3.43E-02 kg 1,4-DCB, 3.00E-02 kg 1,4-DCB, and 2.19E+00 m^2 annual crop eq, respectively. While in the present study, in the same categories the operation presented the values 8.94E-01 kg 1,4-DCB, 1.23E+00 kg 1,4-DCB, and 7.97E-02 m^2 annual crop eq, respectively; higher values that could be reduced if the process is modified.

The second perspective refers to the use of residual by biological processing, Sc 3 or Sc 4; eliminating the drying operation would reduce between 36% and 47% of the global value tested in all categories of Sc 3 and between 42% and 51% in Sc 4. For example, for Sc 3, removing the drying operation would represent a reduction in the overall value, in the categories with the greatest impact, HnCT, GW, WC, TE, and FRS, from 36.92 kg 1,4-DCB, 185.34 kg CO_2 eq, 2001.95 m3, 270.68 kg 1,4-DCB, and 68.29 kg oil eq, respectively, to a value equal to 20.93 kg 1,4-DCB, 108.10 kg CO_2 eq, 1073.01 m^3, 151.52 kg 1,4-DCB, and 42.55 kg oil eq.

Now, if both strategies for reducing environmental impact are considered, the most beneficial would be Sc 5 because it involves extraction processes such as Sc 2 and drying of the initial process, either because it is part of the limit defined for Sc 5 or because the EEC derived from Sc 4 is used. In terms of all categories, the global value tested would be reduced from 27% to 52% and specifically in the IR, WC, and MEu categories, which represent the highest values, the reduction would be 46.7%, 52%, and 44.8%, respectively.

The practice of making these rethinking scenarios is required when you want to convince that the option for a sustainable economy is in the substitution of processes or reagents, such is the case of the study carried out by De Marco et al. (2018) where the agricultural and extraction processes showed the most significant impacts, for which they proposed reducing the use of fertilizers, applying more traditional farming practices and the use of alternative electrical sources, achieving a 17.6% reduction in the impact on categories with damage to human health, 10.3% in ecosystem diversity, and 16.1% in resource availability.

Finally, considering the products generated in the five proposed scenarios entails evaluating their cost in the market, as well as their production method, and this study was limited to the exploration of possible products to be generated in this biorefinery; therefore, thanks to the information derived from the LCA, it is possible to propose the least polluting processing considering the potential of its product, as well as possible improvements toward reducing the environmental impact.

3.5 Conclusions

Life cycle analysis can be a powerful tool for decision-making regarding the environmental impact of biotechnological processes. Defining scenarios that are nothing more than the byproducts to be obtained while refining biological material is essential to apply LCA to a biorefinery successfully. In the case of the coffee waste biorefinery, five scenarios were identified. After the respective analysis, it was determined that the scenario with the lowest environmental impact involved the least energy use and chemical reagents. It is important to note that greater precision is needed for each process and to consider a greater number of scenarios for a complete use of the raw material so that the coffee processing integrates the use of its waste to generate an industry with zero environmental impact.

References

Alzate-Tamayo, L., Jimenez-Cartagena, C., & Londoño-Londoño, J. (2011). Aprovechamiento de residuos agroindustriales para mejorar la calidad sensorial y nutricional de productos avícolas. *Producción + Limpia, 6*(1), 108–127.

Aristizábal-Marulanda, V., Chacón-Perez, Y., & Cardona Alzate, C. A. (2017). The biorefinery concept for the industrial valorization of coffee processing by-products. In *Handbook of coffee processing by-products: Sustainable applications*. Elsevier. https://doi.org/10.1016/B978-0-12-811290-8.00003-7

Ballesteros, L. F., Teixeira, J. A., & Mussatto, S. I. (2014). Chemical, functional, and structural properties of spent coffee grounds and coffee silverskin. *Food and Bioprocess Technology, 7*(12), 3493–3503. https://doi.org/10.1007/s11947-014-1349-z

Battista, F., Zanzoni, S., Strazzera, G., Andreolli, M., & Bolzonella, D. (2020). The cascade biorefinery approach for the valorization of the spent coffee grounds. *Renewable Energy, 157*, 1203–1211. https://doi.org/10.1016/j.renene.2020.05.113

Bichyova, E. E., Sotnikova, L. N., & Maslova, I. N. (2019). Investment development of agroindustrial complex based on the demands of environmental safety. *IOP Conference Series: Earth and Environmental Science, 341*(1). https://doi.org/10.1088/1755-1315/341/1/012213

CADIS, (2011). SimaPro. Available at: https://www.simapro.mx

Caldeira, C., Vlysidis, A., Fiore, G., De Laurentiis, V., Vignali, G., & Sala, S. (2020). Sustainability of food waste biorefinery: A review on valorisation pathways, techno-economic constraints, and environmental assessment. *Bioresource Technology, 312*, 123575. https://doi.org/10.1016/j.biortech.2020.123575

Carrillo-González, G., & Ponce-Sánchez, J. I. (2019). Economía circular, bioeconomía y biorrefinerías. In G. Carrillo Gonzalez & L. G. Torres Bustillos (Eds.), *Biorrefinerías y economía circular* (pp. 15–36).

Catalán, E., Komilis, D., & Sánchez, A. (2019). Environmental impact of cellulase production from coffee husks by solid-state fermentation: A life-cycle assessment. *Journal of Cleaner Production, 233*, 954–962. https://doi.org/10.1016/j.jclepro.2019.06.100

Cherubini, F., Jungmeier, G., Wellisch, M., Willke, T., Skiadas, I., van Ree, R., & de Jong, E. (2009). Toward a common classification approach for biorefinery systems. *Biofuels, Bioproducts and Biorefining, 6*(3), 534–546. https://doi.org/10.1002/bbb.172

Cury R.K., Aguas M.Y., Martinez M.A., Olivero V.R., y Chams C.L. (2017). Residuos agroindustriales su impacto, manejo y aprovechamiento. Revista Colombiana de Ciencia Animal-RECIA, 9(S), 122. https://doi.org/https://doi.org/10.24188/recia.v9.ns.2017.530

De Marco, I., Riemma, S., & Iannone, R. (2018). Life cycle assessment of supercritical CO2 extraction of caffeine from coffee beans. *Journal of Supercritical Fluids, 133*, 393–400. https://doi.org/10.1016/j.supflu.2017.11.005

Farah, A., & Dos Santos, T. F. (2015). The coffee plant and beans: An introduction. In *Coffee in health and disease prevention* (pp. 5–10). Elsevier Inc. https://doi.org/10.1016/B978-0-12-409517-5.00001-2

Gnansounou, E., & Pandey, A. (2017). Classification of biorefineries taking into account sustainability potentials and flexibility. In *Life-cycle assessment of biorefineries* (pp. 1–39). Elsevier B.V. https://doi.org/10.1016/B978-0-444-63585-3.00001-2

Gomez-Millan, G. (2015). Desarrollo de biorrefinerías en el mundo (Biorefineries development: A worldwide review). *Ciencia y Desarrollo*, 34–57.

Green Delta. (2023). *openLCA Nexus*. https://nexus.openlca.org/databases

Hikichi, S. E., Andrade, R. P., Dias, E. S., & Duarte, W. F. (2017). Biotechnological applications of coffee processing by-products. In *Handbook of coffee processing by-products: Sustainable applications* (pp. 221–244). Elsevier. https://doi.org/10.1016/B978-0-12-811290-8.00008-6

Huijbregts, M. A. J., Steinmann, Z. J. N., Elshout, P. M. F., Stam, G., Verones, F., Vieira, M., Zijp, M., Hollander, A., & van Zelm, R. (2016). ReCiPe2016: A harmonised life cycle impact assessment method at midpoint and endpoint level. *The International Journal of Life Cycle Assessment, 22*(2). https://doi.org/10.1007/s11367-016-1246-y

INEGI. (2014). Censo Nacional de Gobiernos Municipales y Delegacionales.

International Coffee Organization. (2020). Producción total por países exportadores. In International Coffee Organization. https://www.ico.org/historical/1990onwards/PDF/1a-total-production.pdf

ISO. (2006). ISO 14040:2006(es). Gestión ambiental-Análisis del ciclo de vida-Principios y marco de referencia. https://www.iso.org/obp/ui#iso:std:iso:14040:ed-2:v1:es:fn:1

Janissen, B., & Huynh, T. (2018). Chemical composition and value-adding applications of coffee industry by-products: A review. *Resources, Conservation and Recycling, 128*, 110–117. https://doi.org/10.1016/j.resconrec.2017.10.001

Kamm, B., Kamm, M., Gruber, P. R., & Kromus, S. (2008). Biorefinery systems: An overview. In *biorefineries-industrial processes and products: Status quo and future directions* (Vol. 1). https://doi.org/10.1002/9783527619849.ch1

López-Arenas, T., & Anaya-Reza, O. (2019). Criterios de evaluación técnico-económico-ambiental para la selección de productos y el diseño de biorrefinerías. In G. Carrillo Gonzalez & L. G. Torres Bustillos (Eds.), *Biorrefinerías y economía circular* (pp. 79–120).

Murthy, P. S., & Madhava, M. (2012). Sustainable management of coffee industry by-products and value addition – A review. *Resources, Conservation and Recycling, 66*, 45–58. https://doi.org/10.1016/j.resconrec.2012.06.005

Mussatto, S. I., Machado, E. M. S., Martins, S., & Teixeira, J. A. (2011). Production, composition, and application of coffee and its industrial residues. *Food and Bioprocess Technology, 4*(5), 661–672. https://doi.org/10.1007/s11947-011-0565-z

OECD. (2006). In M. Oborne (Ed.), *The bioeconomy to 2030: Designing a policy agenda*. OECD.

Oliveira, L. S., & Franca, A. S. (2015). An overview of the potential uses for coffee husks. In *Coffee in health and disease prevention* (pp. 283–291). Elsevier Inc. https://doi.org/10.1016/B978-0-12-409517-5.00031-0

ONU. (2018). La Agenda 2030 y los Objetivos de Desarrollo Sostenible: una oportunidad para América Latina y el Caribe. In Publicación de las Naciones Unidas. https://repositorio.cepal.org/bitstream/handle/11362/40155/24/S1801141_es.pdf

Pandey, A., Soccol, C. R., Nigam, P., Brand, D., Mohan, R., & Roussos, S. (2000). Biotechnological potential of coffee pulp and coffee husk for bioprocesses. *Biochemical Engineering Journal, 6*(2), 153–162. https://doi.org/10.1016/S1369-703X(00)00084-X

Rajesh Banu, J., Kavitha, S., Yukesh Kannah, R., Dinesh Kumar, M., Preethi, Atabani, A. E., & Kumar, G. (2020). Biorefinery of spent coffee grounds waste: Viable pathway towards circular bioeconomy. *Bioresource Technology, 302*, 122821. https://doi.org/10.1016/j.biortech.2020.122821

Sánchez, D. A., & Anzola, C. (2012). Caracterización química de la película plateada del café (*Coffea arabica*) en variedades colombia y caturra. *Revista Colombiana de Química, 41*(2), 211–226.

Schmidt-Rivera, X. C., Gallego-Schmid, A., Najdanovic-Visak, V., & Azapagic, A. (2020). Life cycle environmental sustainability of valorisation routes for spent coffee grounds: From waste to resources. *Resources, Conservation and Recycling, 157*, 104751. https://doi.org/10.1016/j.resconrec.2020.104751

Solarte-Toro, J. C., & Cardona Alzate, C. A. (2021). Biorefineries as the base for accomplishing the sustainable development goals (SDGs) and the transition to bioeconomy: Technical aspects, challenges and perspectives. *Bioresource Technology, 340*, 125626. https://doi.org/10.1016/j.biortech.2021.125626

Wernet, G., Bauer, C., Steubing, B., Reinhard, J., Moreno-Ruiz, E., & Weidema, B. (2016). The ecoinvent database version 3 (part I): Overview and methodology. *International Journal of Life Cycle Assessment, 21*(9), 1218–1230. https://doi.org/10.1007/s11367-016-1087-8

Yepes, M. S., Naranjo, M., Johana, L., & Sánchez Orozco, F. (2008). Valorización de residuos agroindustriales –frutas – en Medellín y el Sur del Valle del Aburrá, Colombia. *Revista Facultad Nacional de Agronomía de Medellín, 61*(611), 4422–4431. http://www.redalyc.org/articulo.oa?id=179914077018

Zuniga-González, C. A. (2024). Productivity growth and technical efficiency changes in Rice farms: An input-oriented study using biosolid. In *Handbook on data envelopment analysis in business, finance, and sustainability: Recent trends and developments* (pp. 551–574). https://doi.org/10.1142/9781800615786_0019

Zuniga-González, C. A., & Jaramillo-Villanueva, J. L. (2022). Frontier model of the environmental inefficiency effects on livestock bioeconomy. *F1000Research, 11*, 1382. https://doi.org/10.12688/f1000research.128071.3

Chapter 4
Valorization of Agro-industrial Residues by the Recovery of Functional Compounds by Green Technologies: Green Tomato Waste, Pumpkin Residues, and Apple Bagasse as Examples from Mexico

Gerardo Valles-Salas, Juliana Morales-Castro,
Víctor Eduardo Cruz-Cárdenas, and Blanca Elizabeth Morales-Contreras

Abstract The circular bioeconomy encompasses two key approaches: the bioeconomy, which focuses on using renewable biological resources to produce sustainable goods, and the circular economy, which aims to minimize waste generation through various methods, such as resource recovery in production cycles. The generation of agro-industrial waste is a global issue with critical implications due to its alarming increase in production rates. This highlights the importance of developing strategies for sustainable valorization. Among these, the recovery of bioactive and techno-functional compounds, particularly for health benefits, stands out as a high-potential approach. Emerging extraction techniques play a fundamental role in improving the efficiency and quality of compound recovery. Notable cases include the extraction of bioactive pectin from green tomato peels (tomatillo), the recovery of carotenoids using green solvents from pumpkin seed industry waste, and the valorization of pectin from apple pomace. These examples highlight innovative strategies and ongoing challenges in leveraging agro-industrial waste as a resource, fostering sustainable practices within the framework of the circular bioeconomy.

Keywords Agro-industrial waste · Bioactive compound · Circular bioeconomy · Sustainable

---

G. Valles-Salas · J. Morales-Castro (✉) · V. E. Cruz-Cárdenas · B. E. Morales-Contreras
Departamento de Ingenierías Química y Bioquímica, Tecnológico Nacional de México/Instituto Tecnológico de Durango. LANAEPBI, Durango, Mexico
e-mail: 13040842@itdurango.edu.mx; jmorales@itdurango.edu.mx; 23041389@itdurango.edu.mx; bmorales@itdurango.edu.mx

S. G. Ceballos Pérez et al. (eds.), *Sustainability in Latin America*,
https://doi.org/10.1007/978-3-032-09787-3_4

4.1 Introduction

Circular economy is defined as a model that focuses on reducing, reusing, and recycling materials to minimize waste and promote the regeneration of sustainable systems. In this sense, the integration of circular economy models and tools at various stages of the food supply chain is necessary to create a closed-loop agri-food system that enables the valorization of generated waste and consequently reduces its generation (Diéguez-Santana et al., 2022).

In the agri-food industry, raw materials are subjected to adaptation or transformation processes through a series of unit operations that facilitate their consumption. However, these processes lead to a high production of waste (Cury et al., 2017). Major concerns regarding the generation of these wastes include their contribution to environmental pollution due to their high biodegradability, along with the economic losses they cause and the social concerns they generate (Martínez-Inda et al., 2023). Therefore, it is necessary to implement a significant change in the agri-food system to manage these wastes effectively. In this context, the circular economy model offers an efficient solution by promoting the reuse, valorization, and recovery of natural resources from by-products of this industry (Osorio et al., 2021).

The valorization of agro-industrial waste through the circular economy model aims to reintroduce waste into a new process as raw material to produce valuable bio-functional compounds with applications in the food, health, pharmaceutical, cosmetic, and environmental sectors, adding value to various products through sustainable technology (Martínez-Inda et al., 2023). These compounds can replace synthetic preservatives, pigments, fragrances, and antioxidants in food and cosmetic products, as well as provide protective health benefits (Berenguer et al., 2023).

Husk tomato (tomatillo) waste, apple pomace, and pumpkin waste have been identified as potential sources for the extraction of bioactive and techno-functional compounds. These wastes have been valorized using green solvents and emerging technologies, under the circular economy approach (Morales-Contreras et al., 2017, 2020; Portillo-López et al., 2021).

The objective of this chapter is to analyze strategies and technologies for the valorization of agri-industrial waste, with an emphasis on the recovery of functional compounds. Case studies are examined, and the opportunities and challenges associated with these practices are explored, to promote a circular and sustainable economy in the region.

4.2 Circular Bioeconomy in Waste Valorization Explores Various Strategies and Technologies for Transforming Waste

4.2.1 Bioeconomy and Circular Economy

The bioeconomy is based on an economic model in which the production of goods and services is mainly based on the efficient and sustainable use of biological resources (genes, biomass, bacteria, plants, and animals) and natural resources (water and soil) in addition to the use of waste and residues generated during their transformation, with the aim of reducing the use of fossil energy and contributing to a global objective of decarbonizing the economy (De Jaramillo et al., 2019; Hernández & Céspedes, 2020). Likewise, the circular bioeconomy has a great impact on the valorization of agro-industrial waste due to its main approach which is based on the reduction of waste and/or adding that waste to a production chain thus keeping the cycle closed, through reuse, recycling, and valorization of resources (Yepes et al., 2008).

The bioeconomy arises in response to four challenges: (1) the increase in population worldwide and food security, (2) the great increase in waste without a value, (3) the change of fossil fuels for alternative energies, and (4) the great concern about climate change. Therefore, the main axis of the bioeconomy are natural resources, biological principles and processes, as well as technologies that are intertwined with knowledge, development, transformation, or regeneration (De Jaramillo et al., 2019).

The fusion of bioeconomy and circular economy concepts into circular bioeconomy (CBE) has emerged as a sustainable economic model, addressing economic, social, and environmental dimensions (Hernández & Céspedes, 2020). CBE integrates key elements from both frameworks, including improved resource use, eco-efficiency, carbon footprint reduction, and waste valorization; CBE is applied in biomass-based sectors, where circularity principles are implemented sequentially throughout production processes (Borrego-Marín et al., 2021).

4.2.2 Valorization of Agri-food Waste by a Circular Bioeconomy Approach

The circular bioeconomy seeks to replace the linear economic model with a more sustainable and responsible approach, emphasizing resource optimization and waste reduction. This model leverages agro-industrial waste, aligning with the principles of sustainability (Gómez-Ramos, 2023).

Agro-industrial residues, including leaves, roots, straw, bagasse, and husks, are abundant by-products of human activities like agriculture and livestock. In many Latin American countries, these wastes are not adequately regulated for treatment before disposal. Although waste management strategies have been proposed, their implementation faces obstacles, often leading to harmful practices such as burning or landfilling, which pose economic, environmental, and social issues (Da Freiria Ferreira et al., 2024).

Given their composition, these residues can be utilized as raw materials for energy generation and for the recovery of valuable bioactive and techno-functional compounds (Martínez-Inda et al., 2023). Their integral utilization, combined with the search for alternative sources for bioactive compound extraction, offers an innovative strategy that enhances sustainability and resource efficiency.

Ultimately, this strategy merges the principles of the circular economy and bioeconomy, transforming waste into high-value products through eco-friendly methods that benefit human health (Díaz-Vázquez et al., 2021).

4.3 Use of Agro-industrial Wastes in the Recovery of Functional Compounds

Foods contain nutrients, undesirable compounds, and bioactive compounds, the latter offering health benefits beyond nutrition (Teodoro, 2019). These bioactive remain in food matrices, even in agro-industrial waste. Adopting circular bioeconomy processes is essential to recover these compounds, minimizing economic loss and maximizing their potential in food and pharmaceutical industries (Lemes et al., 2022).

4.3.1 Bioactive and Techno-functional Compounds

Bioactive compounds, primarily produced by plants as secondary metabolites, play roles in competition, defense, attraction, and signaling (Leichtweis et al., 2021). These compounds, while generally lacking nutritional value, are naturally present in foods and exhibit biological activity that can modulate metabolic processes to enhance health (Câmara et al., 2021). Their benefits include antioxidant, antimicrobial, anticancer, and anti-inflammatory properties (García-Mahecha et al., 2023).

Bioactive compounds can be broadly classified into several groups, including polyphenols, terpenoids, phytosterols, polysaccharides, glucosinolates, alkaloids, capsaicinoids, carotenoids, and tocopherols (de Freitas Marinho et al., 2023). Among these, polyphenols are especially common and are divided into flavonoids (such as anthocyanins, flavanols, flavanones, flavanols, flavones, and isoflavones) and non-flavonoid polyphenols (such as phenolic acids, xanthones, stilbenes, lignans, and tannins). These compounds have potent antioxidant properties and have been linked to reduced cardiovascular disease risk and improved well-being (Durazzo et al., 2019).

Polysaccharides, including cellulose, pectin, starch, chitin, hyaluronic acid, and inulin, are vital for gut microbiota and overall health (de Souza et al., 2022). Glucosinolates, found in vegetables like broccoli and cabbage, contribute to their distinctive taste and odor, with bioactivities such as chemo preventive and anti-inflammatory effects. Capsaicin, the main compound in capsaicinoids from chili peppers, provides a burning sensation and offers anti-inflammatory, antioxidant, and anti-obesity effects, with potential anticancer and antimutagenic properties (Cortés-Ferré et al., 2021).

Carotenoids (such as α-carotene, β-carotene, lutein, and lycopene) and tocopherols (such as α- and γ-tocopherols) are precursors to vitamins A and E, essential in the human diet (Viñas-Ospino et al., 2023).

Additionally, some of these compounds, such as polysaccharides, possess techno-functional properties crucial in food processing, such as emulsification, solubility, water and oil binding capacity, foam capacity and stability, gelation, and viscosity (Caldeira et al., 2024). For example, pectin benefits gut microbiota while serving as a stabilizing and thickening agent in foods.

4.3.2 Valorization of Agro-industrial Waste in the Recovery of Compounds

Agro-industrial residues are rich in bioactive compounds but present extraction challenges due to their complex, heterogeneous nature and pH variations. These compounds are often sensitive to degradation by oxygen, complicating recovery (Rodríguez et al., 2023). Extraction processes aim to maximize yield and optimize conditions while minimizing degradation, costs, and environmental impact (Leichtweis et al., 2021). Methods include conventional techniques (Soxhlet extraction maceration, percolation, and hydrodistillation) and emerging technologies such as microwave-assisted, ultrasound-assisted extraction, and supercritical fluid extraction (Ampese et al., 2024).

4.3.3 Extraction Methods of Bioactive Compounds

Solid-liquid extraction is commonly used to recover bioactive compounds from agro-industrial residues by mixing a solid material with a solvent to separate and purify the target compound (Leichtweis et al., 2021). Conventional methods, such as Soxhlet extraction, involve solvent evaporation and repeated interaction with the solid. Other techniques include maceration, which employs continuous agitation, percolation, where the solvent slowly passes through the solid, and hydrodistillation (Carvajal Quiceno, 2023). However, these methods often have drawbacks, such as long extraction times, high solvent consumption, and significant energy use

(Gonçalves Rodrigues et al., 2019). Emerging techniques like ultrasound-assisted extraction, microwave-assisted extraction, and supercritical fluid extraction offer enhanced efficiency and reduced environmental impact (Campo-Vera et al., 2023; Otálora-Orrego & Martin G., 2020). These innovations provide more sustainable alternatives for industrial applications.

Ultrasound-assisted extraction (UAE) utilizes cavitation, where ultrasonic waves create bubbles that expand and collapse, damaging cell walls and enhancing compound recovery. Key factors include penetration into cell walls and vibration-induced cell damage (Benavides-Guerrero et al., 2020). UAE is efficient, is economical, and reduces energy and solvent use. However, it can generate heat that may degrade heat-sensitive compounds, and method reproducibility may decline with instrument aging (Martínez-Zamora et al., 2024).

Microwave-assisted extraction consists in the use of electromagnetic waves damaging the cell wall, allowing a higher mass transfer, involving two phenomena: (1) ionic conductivity: resistance to the flow of electricity posed by the cells and the rotation of the dipoles, (2) extremely high oscillation frequency within the electromagnetic fields causing collisions between the dipoles trying to align with the field direction leading to friction and in the generation of heat (Cruz et al., 2019; Garcia, 2017). It is a cost-effective device that significantly reduces extraction time and solvent usage while enhancing extraction efficiency.

Supercritical fluid extraction (SFE) is a method that relies on the use of a solvent fluid in its supercritical state. To achieve this, temperature and pressure parameters are controlled to reach conditions where the fluid is between the gas and liquid states, displaying liquid density and gas-like viscosity. Carbon dioxide (CO_2) is an attractive and widely used solvent in this technique (da Silva et al., 2016). This method offers high selectivity for non-polar compounds, does not degrade heat-sensitive compounds, and leaves no toxic solvent residues. However, it involves high costs and requires complex operation and training (García-Pérez et al., 2017).

4.4 Application of Agro-industrial Waste Valorization in México

Studies in Ibero-America highlight the success of agro-industrial waste valorization, positioning the region as a leader in bioeconomy and sustainability. Valorization transforms waste into value-added products, enhancing resource efficiency and reducing reliance on non-renewable materials. This approach offers sustainable, economically viable solutions for waste management, especially in sectors like food, pharmaceuticals, cosmetics, and the environment. Examples from Mexico, such as the utilization of apple pomace, tomatillo husks, and pumpkin residues, demonstrate how circular economy practices can drive innovation and sustainability in the agro-industry.

4.4.1 *Case Study 1: Evaluation of Husk Green Tomato or Tomatillo (By-products) for Pectin Recovery and Bioactivity Potential*

Pectin is a polysaccharide renowned for its gelling properties, widely used in the food industry, particularly in confectionery products. It also plays essential biological roles, including hydration, swelling, thickening, and gel formation (Gerschenson et al., 2021). However, the primary sources of industrial pectin citrus fruits and apples are insufficient to meet the growing market demand driven by applications in the cosmetic, pharmaceutical, and food industries (Santos et al., 2020). Identifying new pectin sources presents valuable opportunities and aligns with bioeconomy principles by valorizing agro-industrial waste.

The "tomatillo," green husk tomato, *Physalis ixocarpa* broth a plant native to Mexico, produces a fruit known for its distinctive flavor both raw and cooked, and it is widely used in sauces in Latin American cuisine. Additionally, due to inadequate storage, overproduction, and its short shelf life, significant amounts of waste are generated. This waste has been investigated for its potential to yield functional compounds, specifically polysaccharides like pectin (Fig. 4.1), as indicated by the studies of Morales-Contreras et al. (2017).

In one study, pectin was obtained from discarded husk tomato achieving a pectin yield of 19.8%. The properties found, such as the molecular weight ranging from 542 to 699 kDa, viscosity with pseudoplastic behavior, the presence of neutral sugars (fucose, rhamnose, arabinose, galactose, glucose, and xylose), and a galacturonic acid content of 60%, demonstrate that the pectin obtained is comparable to traditional sources. Moreover, the neutral sugar composition suggests that this

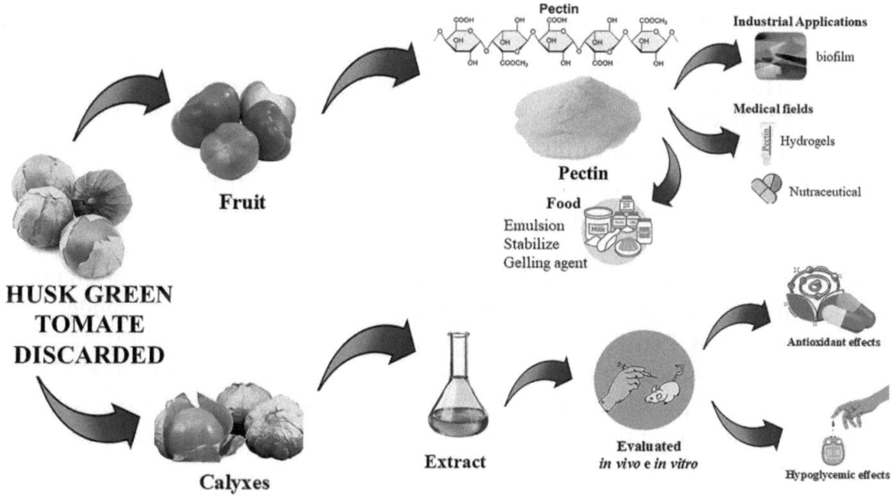

Fig. 4.1 Valorization process of husk green tomato or tomatillo obtaining pectin

pectin has a wide range of applications in the food and pharmaceutical industries. This offers an alternative to conventional pectin sources (Morales-Contreras, 2017). Further, the functionality of this high methoxyl pectin in gel formation using citric acid and HCl was evaluated by means of rheological studies and textural attributes that indicate thermostable husk tomato pectin gels that could be used in the design and control of gel microstructure where non-thermo-reversible gels are required to sustain processing at higher temperatures. The utilization of these pectin gels to improve textural attributes was corroborated from penetration studies where the rupture strength was higher or comparable to values from gels from citrus pectins (Morales-Contreras et al., 2018).

In addition, the hypoglycemic and antioxidant potential of discarded husks (calyces) from *P. ixocarpa* variety was evaluated (Guerrero-Romero et al., 2021). Extracts obtained through maceration, infusion, and ultrasound-assisted extraction were assessed in vitro by analyzing glucose diffusion and the inhibitory action on α-amylase and α-glucosidase enzymes, as well as in vivo using an animal model. The findings showed that these extracts could enhance glucose metabolism, suggesting their potential as a natural remedy in hypoglycemic therapy for diabetic patients.

As such has been described, agro-industrial waste from tomatillo processing can be utilized to extract valuable components such as pectin and its potential use in the food industry and being the base for nutraceutical product development in addition to reducing waste and contributing to the circular bioeconomy.

4.4.2 Case Study 2: Valorization of Pumpkin Waste from the Seed Industry for the Extraction of Pigments

In Mexico, the states of Zacatecas, Campeche, Guerrero, Michoacán, and Tamaulipas have a seed industry dedicated to cultivating and harvesting pumpkins solely for their seeds (Lorenzo-Hernández et al., 2019); consequently, the rest of the fruit (pulp and peel) becomes waste. These residues account for 79–82% and 10–12% of the total weight of the pumpkin fruit, respectively (Portillo-López et al., 2021). Due to the lack of proper management, these wastes are either discarded or left in the fields to decompose and occasionally used as unconventional feed for livestock and poultry.

Therefore, it is essential to explore alternatives for the valorization of pumpkin seed residues that enable cycle closure through the circular economy model. The study conducted by Portillo-López et al. (2021) revealed that pumpkin waste from Zacatecas has significant potential as a source of natural pigments, specifically carotenoids (Fig. 4.2). Additionally, the study investigated the use of green solvents as alternatives to conventional ones, thus supporting the development of an ecological process.

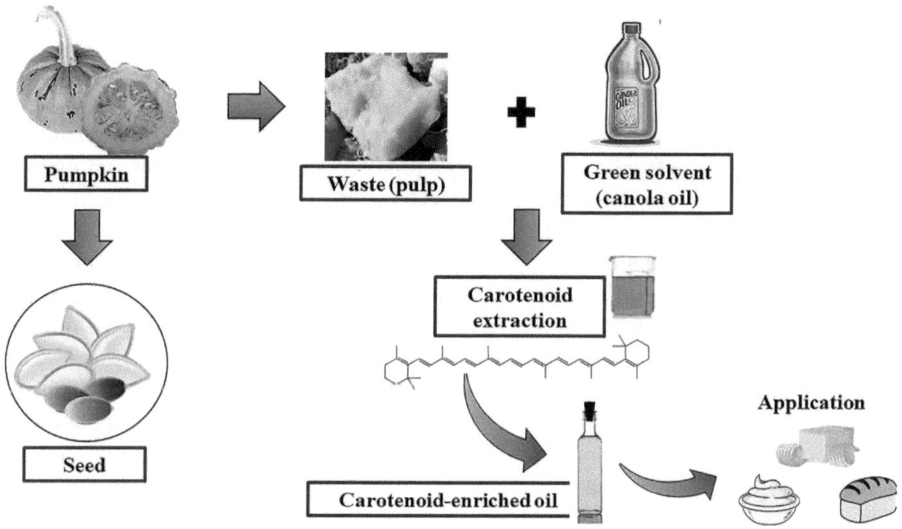

Fig. 4.2 Pumpkin waste valorization process to obtain carotenoids using canola oil as green solvent

In this study, vegetable oils from canola, corn, and soybean were used as green and ecological alternatives to conventional organic solvents for extracting carotenoids from dehydrated pumpkin pulp waste. Various extraction conditions, such as time, agitation speed, and temperature, were investigated, establishing optimal parameters of 21.8 min, 250 rpm, and 60 °C. Under these conditions, an extraction yield of 378.1 μg of β-carotene equivalents per gram of sample was achieved, representing 61.6% of the total carotenoid content in the pumpkin pulp waste.

Through this extraction methodology, an effective valorization of the waste was achieved, resulting in a value-added product: a commercial oil naturally enriched with carotenoids. This oil could be used as a functional ingredient in the production of food items (such as bakery products, margarine, mayonnaise, and dressings), cosmetics (body and skin care products), and pharmaceuticals (nutraceuticals). The high potential of carotenoids, with their provitamin A properties and antioxidant activity, enhances the value of the oils.

This approach allows the utilization of an agro-industrial by-product, promotes waste reduction, and ensures the integral use of biomass. Moreover, it results in value-added products, such as commercial oils enriched with compounds that provide health benefits. This is crucial in the bioeconomy, where the goal is to maximize the sustainable use of biological resources and improve the quality of life for humans.

4.4.3 Case Study 3: Valorization of Apple Pomace as a Sustainable Source of Pectin

Apples are among the most widely cultivated and consumed fruits globally, significantly impacting the food industry through products like juice and cider. However, the processing of apples for these products results in millions of tons of apple pomace annually (Ampese et al., 2024). Managing this by-product effectively is crucial, presenting a promising opportunity for waste valorization in the agro-food industry (Costa et al., 2022). This aligns with the circular economy concept, which focuses on recycling waste to improve economic efficiency and support sustainable development.

Apple pomace, a major source of pectin alongside citrus peels and sugar beets (Canteri-Schemin et al., 2005), can be valorized to extract bioactive compounds. In Durango, Mexico, the "Blanca de Asturias" apple variety, used for pollination of Golden and Red Delicious apples, is processed into products like jellies and jams. This process generates substantial pomace waste, which is also used for animal feed. Therefore, valorizing apple pomace to produce high-value products through circular economy processes is essential for enhancing sustainability and resource efficiency.

A study by Morales-Contreras et al. (2020) evaluated the potential of "Blanca de Asturias" apple pomace as a sustainable pectin source (Fig. 4.3). Proximal analysis determined the total pectin content in the apple pomace to be 33.5% ± 4.1, similar to pectin levels found in citrus peels. Thus, this agro-industrial waste can be considered a potential pectin source.

The valorization of apple pomace through a circular bioeconomy approach was conducted by using the residue from apple juice production as raw material for pectin extraction. Pectin extraction yield of 36.0 mg per gram of pomace was

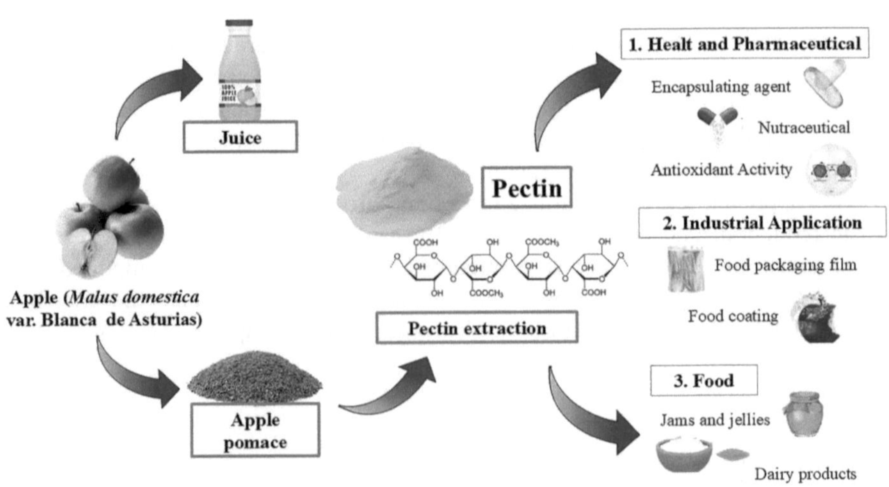

Fig. 4.3 Valorization of apple pomace to obtain pectin and its various applications

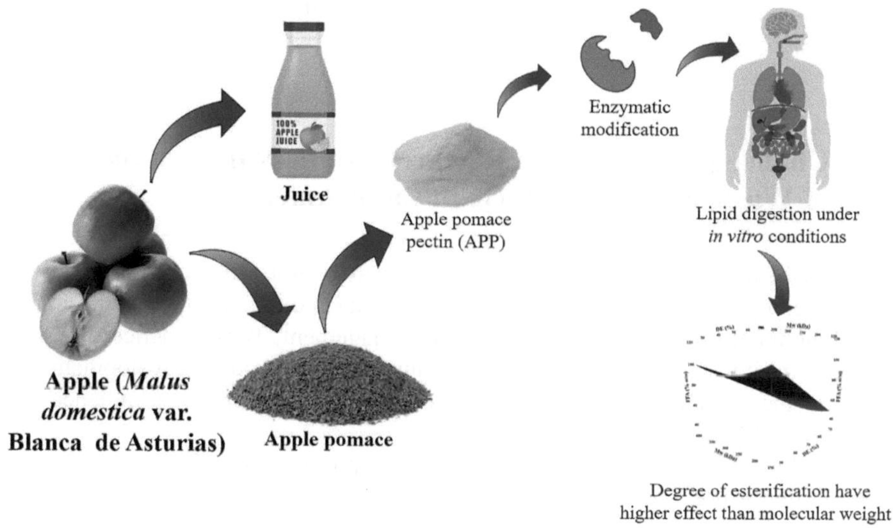

Fig. 4.4 Application of pectin extracted from apple bagasse in lipid digestibility

obtained. The physicochemical properties of pectin indicated a galacturonic acid content ranging from 38.0 to 47.1 g per 100 g of sample, the presence of neutral sugars like glucose, galactose, arabinose, xylose, and mannose, a molecular weight between 865 kDa and 590 kDa, and rheological properties including flow behavior, activation energy, viscosity, and gelling point, similar to those from commercial pectin, primarily derived from citrus. Moreover, due to its gelling, stabilizing, and thickening properties, the extracted pectin has potential applications in the food industry for making jams, jellies, yogurts, and dairy products and pharmaceutical industry, and it can be used in formulating medications and dietary supplements due to its digestive health benefits and in cosmetics for producing gels and emulsions (Morales-Contreras et al., 2020).

Moreover, derived from these properties, a further study investigated the use of pectin derived from apple pomace (APP) in lipid digestion. Through enzymatic de-esterification, modified apple pectin (MAPP) with varying degrees of esterification and molecular weights was obtained. Both APP and MAPP were incorporated into oil-in-water emulsions and their lipid digestibility was assessed using a multistage in vitro model (Fig. 4.4). The results revealed that MAPP reduced lipid digestibility by 25%, indicating its potential in the formulation of high-fat foods that could aid in addressing obesity (Morales-Contreras et al., 2024). This exemplifies the valorization of apple pomace within a circular economy, transforming waste into high-value ingredients for the nutraceutical and food industries.

This valorization strategy for apple pomace in pectin recovery exemplifies how integrating circular economy principles can transform agro-industrial waste management. This approach promotes a more sustainable and diversified economy by reducing the amount of waste generated and opens new economic opportunities by converting waste into high-value ingredients for the nutraceutical and food industries.

4.5 Challenges and Future Research on the Valorization of Waste and By-products in Ibero-America

Based on the reviewed research, agro-industrial waste valorization can play a crucial role in the bioeconomy and contribute to the circular economy, offering sustainable and economically viable solutions for waste management. These practices not only reduce agro-industrial waste but also drive innovation in product development and enhance the efficiency of biological resource use, decreasing reliance on non-renewable raw materials.

Ibero-America shows strong potential for adopting circular bioeconomy practices, turning waste into valuable products. However, achieving this requires developing sustainable extraction technologies and green solvents to recover compounds without losing their bio-functionality as well as the use of green solvents and extraction methods that improve efficiency in the recovery of valuable compounds without compromising their integrity or bio-functionality, allowing their application in various sectors.

It is important to consider the scalability, the challenge lies in scaling these processes for industrial application while maintaining efficiency and sustainability. Successfully implementing these practices could significantly contribute to sustainable development in Ibero-America, moving toward zero-waste processes and maximizing resource use.

4.6 Conclusions

Agro-industrial wastes have generated a great area of opportunity; likewise, it is of utmost importance to incorporate these wastes to the circular bioeconomy model, attaching them to valorize them, decreasing environmental, economic, and social problems, as well as the importance of the recovery of bioactive and techno-functional compounds for a great application in different areas and industries such as food, pharmaceutical and cosmetology, as well as the extraction of these compounds accompanied by unconventional methodologies, green solvents obtaining higher yields, and better quality compounds. The extraction of these compounds is accompanied by non-conventional methodologies, green solvents obtaining a higher yield, and better quality compounds, and clearly there are challenges and areas of opportunity where we must work, at the same time to reduce the waste generated due to the demand in Latin America.

References

Ampese, L. C., Sganzerla, W. G., Di Domenico Ziero, H., Costa, J. M., Martins, G., & Forster-Carneiro, T. (2024). Valorization of apple pomace for biogas production: A leading anaerobic biorefinery approach for a circular bioeconomy. *Biomass Conversion and Biorefinery, 14*(13), 14843–14857. https://doi.org/10.1007/s13399-022-03534-6

Benavides-Guerrero, R., Revelo-Cuarán, Y. A., Arango-Bedoya, O., Osorio-Mora, O., Benavides-Guerrero, R., Revelo-Cuarán, Y. A., Arango-Bedoya, O., & Osorio-Mora, O. (2020). Extracción asistida con ultrasonido de compuestos fenólicos de dos variedades de papas (Solanum phureja) nativas andinas y evaluación de su actividad antioxidante. *Información tecnológica, 31*(5), 43–50. https://doi.org/10.4067/S0718-07642020000500043

Berenguer, C. V., Andrade, C., Pereira, J. A. M., Perestrelo, R., & Câmara, J. S. (2023). Current challenges in the sustainable valorisation of agri-food wastes: A review. *Processes, 11*(1), Article 1. https://doi.org/10.3390/pr11010020

Borrego-Marín, M. d. M., Berbel, J., Gómez-Limón, J. A., & Villanueva, A. J. (2021). Factores habilitantes y limitantes para el desarrollo empresarial de la Bioeconomía Circular. Análisis empírico para el caso de Andalucía. *C3-BIOECONOMY: Circular and Sustainable Bioeconomy, 2*, Article 2. https://doi.org/10.21071/c3b.vi2.13468

Caldeira, R. F., de Paiva Gouvêa, L., de Lima Azevedo, T., Conte, C., de Sá, D. d. G. C. F., Galdeano, M. C., Felberg, I., Lima, J. R., & Mellinger, C. G. (2024). Processing parameters, techno-functional properties and potential food application of lentil protein concentrate as an ingredient for the plant-based market. *Food Research International, 189*, 114569. https://doi.org/10.1016/j.foodres.2024.114569

Câmara, J. S., Albuquerque, B. R., Aguiar, J., Corrêa, R. C. G., Gonçalves, J. L., Granato, D., Pereira, J. A. M., Barros, L., & Ferreira, I. C. F. R. (2021). Food bioactive compounds and emerging techniques for their extraction: Polyphenols as a case study. *Foods, 10*(1), Article 1. https://doi.org/10.3390/foods10010037

Campo-Vera, Y., Gélvez-Ordoñez, V. M., Orozco-Rozo, A. E., Campo-Vera, Y., Gélvez-Ordoñez, V. M., & Orozco-Rozo, A. E. (2023). Efecto del pretratamiento con ultrasonido sobre la calidad del aceite crudo de palma (Elaeis guineensis). *Información tecnológica, 34*(4), 57–66. https://doi.org/10.4067/S0718-07642023000400057

Canteri-Schemin, M. H., Fertonani, H. C. R., Waszczynskyj, N., & Wosiacki, G. (2005). Extraction of pectin from apple pomace. *Brazilian Archives of Biology and Technology, 48*, 259–266. https://doi.org/10.1590/S1516-89132005000200013

Carvajal Quiceno, A. D. (2023). Detección de metabolitos secundarios presentes en las hojas de (Trichanthera Gigantea) Acanthaceae y su actividad antimicrobiana y antimicótica. *Acta Biológica Colombiana, 28*(1), 118–127. https://doi.org/10.15446/abc.v28n1.89656

Cortés-Ferré, H. E., Guajardo-Flores, D., Romero-De La Vega, G., & Gutierrez-Uribe, J. A. (2021). Recovery of Capsaicinoids and other phytochemicals involved with TRPV-1 receptor to re-valorize chili pepper waste and produce nutraceuticals. *Frontiers in Sustainable Food Systems, 4*. https://doi.org/10.3389/fsufs.2020.588534

Costa, J. M., Ampese, L. C., Ziero, H. D. D., Sganzerla, W. G., & Forster-Carneiro, T. (2022). Apple pomace biorefinery: Integrated approaches for the production of bioenergy, biochemicals, and value-added products – An updated review. *Journal of Environmental Chemical Engineering, 10*(5), 108358. https://doi.org/10.1016/j.jece.2022.108358

Cruz, A. M., Ramírez, A. H., García, V. R., & Reyes, Y. C. (2019). Evaluación del efecto de secado con microondas en los compuestos bioactivos del bagazo de marañón, *6*.

Cury, R. K., Aguas M, Y., Martinez M, A., Olivero V, R., & Chams Ch, L. (2017). Residuos agroindustriales su impacto, manejo y aprovechamiento. *Revista Colombiana de Ciencia Animal – RECIA, 9*(S1), 122–132. https://doi.org/10.24188/recia.v9.nS.2017.530

Da Freiria Ferreira, N., Nicolli Eva Pereira Echevarria, Volken, N. J., Wellyngton Silva De Amorim, De Andrade Guerra, J. B. S. O., & Dias, F. T. (2024). The role of waste valorization in the transition to the circular economy. https://doi.org/10.13140/RG.2.2.33808.42242

da Silva, R. P. F. F., Rocha-Santos, T. A. P., & Duarte, A. C. (2016). Supercritical fluid extraction of bioactive compounds. *TrAC Trends in Analytical Chemistry, 76*, 40–51. https://doi.org/10.1016/j.trac.2015.11.013

de Freitas Marinho, L., Sganzerla, W. G., Ferreira, V. C., Jimenez Moreno, J. A., Rostagno, M. A., & Forster-Carneiro, T. (2023). Advances in green extraction methods, biological properties, and applications of betanin and vitexin: An updated review and bibliometric analysis. *Biocatalysis and Agricultural Biotechnology, 51*, 102744. https://doi.org/10.1016/j.bcab.2023.102744

De Jaramillo, E. H., Henry, G., & Trigo, E. (Eds.). (2019). La bioeconomía. Nuevo marco para el crecimiento sostenible en América Latina: Bioeconomy. New framework for sustainable growth in Latin America. Pontificia Universidad Javeriana. https://doi.org/https://doi.org/10.2307/j.ctvkwnpxt

Díaz-Vázquez, D., Carrillo-Nieves, D., Orozco-Nunnelly, D. A., Senés-Guerrero, C., & Gradilla-Hernández, M. S. (2021). An integrated approach for the assessment of environmental sustainability in agro-industrial waste management practices: The case of the tequila industry. *Frontiers in Environmental Science, 9*. https://doi.org/10.3389/fenvs.2021.682093

Diéguez-Santana, K., Sarduy-Pereira, L. B., Sablón-Cossío, N., Bautista-Santos, H., Sánchez-Galván, F., & Ruíz Cedeño, S. d. M. (2022). Evaluation of the circular economy in a pitahaya agri-food chain. *Sustainability, 14*(5), Article 5. https://doi.org/10.3390/su14052950

Durazzo, A., Lucarini, M., Souto, E. B., Cicala, C., Caiazzo, E., Izzo, A. A., Novellino, E., & Santini, A. (2019). Polyphenols: A concise overview on the chemistry, occurrence, and human health. *Phytotherapy Research, 33*(9), 2221–2243. https://doi.org/10.1002/ptr.6419

Garcia, J. R. (2017). Extracción de compuestos bioactivos mediante pre-tratamiento con microondas de subproductos vitivinícolas. Valorización del raspón de la uva.

García-Mahecha, M., Soto-Valdez, H., Carvajal-Millan, E., Madera-Santana, T. J., Lomelí-Ramírez, M. G., & Colín-Chávez, C. (2023). Bioactive compounds in extracts from the agro-industrial waste of mango. *Molecules, 28*(1), Article 1. https://doi.org/10.3390/molecules28010458

García-Pérez, J. S., Robledo-Padilla, F., Cuellar-Bermudez, S. P., Arévalo-Gallegos, A., Parra-Saldivar, R., Zavala-Yoe, R., Ramirez-Mendoza, R. A., & Iqbal, H. M. N. (2017). Thermodynamics and statistical correlation between supercritical-CO2 fluid extraction and bioactivity profile of locally available Mexican plants extracts. *The Journal of Supercritical Fluids, 122*, 27–34. https://doi.org/10.1016/j.supflu.2016.12.002

Gerschenson, L. N., Fissore, E. N., Rojas, A. M., Idrovo Encalada, A. M., Zukowski, E. F., & Higuera Coelho, R. A. (2021). Pectins obtained by ultrasound from agroindustrial by-products. *Food Hydrocolloids, 118*, 106799. https://doi.org/10.1016/j.foodhyd.2021.106799

Gómez-Ramos, G. (2023). Bioeconomía circular: Aprovechamiento de residuos de frutas y verduras en México. *Mexican journal of technology and engineering, 2*(2), 4–9. https://doi.org/10.61767/mjte.002.2.0409

Gonçalves Rodrigues, L. G., Mazzutti, S., Vitali, L., Micke, G. A., & Ferreira, S. R. S. (2019). Recovery of bioactive phenolic compounds from papaya seeds agroindustrial residue using subcritical water extraction. *Biocatalysis and Agricultural Biotechnology, 22*, 101367. https://doi.org/10.1016/j.bcab.2019.101367

Guerrero-Romero, F., Simental-Mendía, L. E., Guerra Rosas, M. I., Sayago-Monreal, V. I., Morales Castro, J., & Gamboa-Gómez, C. I. (2021). Hypoglycemic and antioxidant effects of green tomato (Physalis ixocarpa Brot.) calyxes' extracts. *Journal of Food Biochemistry, 45*(4). https://doi.org/10.1111/jfbc.13678

Hernández, R. E., & Céspedes, J. (2020). Bioeconomía: Una estrategia de sostenibilidad en la cuarta revolución industrial. *Revista de Investigación e Innovación Agropecuaria y de Recursos Naturales, 7*(2), 126–133.

Leichtweis, M. G., Oliveira, M. B. P. P., Ferreira, I. C. F. R., Pereira, C., & Barros, L. (2021). Sustainable recovery of preservative and bioactive compounds from food industry bioresidues. *Antioxidants, 10*(11), Article 11. https://doi.org/10.3390/antiox10111827

Lemes, A. C., Egea, M. B., de Oliveira Filho, J. G., Gautério, G. V., Ribeiro, B. D., & Coelho, M. A. Z. (2022). Biological approaches for extraction of bioactive compounds from agro-industrial by-products: A review. *Frontiers in Bioengineering and Biotechnology, 9*. https://doi.org/10.3389/fbioe.2021.802543

Lorenzo-Hernández, R., Torres-Salado, N., Sánchez-Santillán, P., Herrera-Pérez, J., Mayrén-Mendoza, F. d. J., Salinas-Ríos, T., Rojas-García, A. R., Maldonado-Peralta, M. d. l. Á., Lorenzo-Hernández, R., Torres-Salado, N., Sánchez-Santillán, P., Herrera-Pérez, J., Mayrén-Mendoza, F. d. J., Salinas-Ríos, T., Rojas-García, A. R., & Maldonado-Peralta, M. d. l. Á. (2019). Evaluación de las características de calidad y bromatológicas de ensilados elaborados con residuos de calabaza (Cucurbita argyrosperma). *Revista Internacional De Contaminación Ambiental, 35*(4), 957–963. https://doi.org/10.20937/rica.2019.35.04.14

Martínez-Inda, B., Esparza, I., Moler, J. A., Jiménez-Moreno, N., & Ancín-Azpilicueta, C. (2023). Valorization of agri-food waste through the extraction of bioactive molecules. Prediction of their sunscreen action. *Journal of Environmental Management, 325*, 116460. https://doi.org/10.1016/j.jenvman.2022.116460

Martínez-Zamora, L., Hashemi, S., Cano-Lamadrid, M., Bueso, M. C., Aguayo, E., Kessler, M., & Artés-Hernández, F. (2024). Ultrasound-assisted extraction of bioactive compounds from broccoli by-products. *Foods, 13*(10), Article 10. https://doi.org/10.3390/foods13101441

Morales-Contreras, B. E., Contreras-Esquivel, J. C., Wicker, L., Ochoa-Martínez, L. A., & Morales-Castro, J. (2017). Husk tomato (Physalis ixocarpa Brot.) waste as a promising source of pectin: Extraction and physicochemical characterization. *Journal of Food Science, 82*(7), 1594–1601. https://doi.org/10.1111/1750-3841.13768

Morales-Contreras, B. E., Rosas-Flores, W., Contreras-Esquivel, J. C., Wicker, L., & Morales-Castro, J. (2018). Pectin from husk tomato (Physalis ixocarpa Brot.): Rheological behavior at different extraction conditions. *Carbohydrate Polymers, 179*, 282–289. https://doi.org/10.1016/j.carbpol.2017.09.097

Morales-Contreras, B. E., Wicker, L., Rosas-Flores, W., Contreras-Esquivel, J. C., Gallegos-Infante, J. A., Reyes-Jaquez, D., & Morales-Castro, J. (2020). Apple pomace from variety "Blanca de Asturias" as sustainable source of pectin: Composition, rheological, and thermal properties. *LWT, 117*, 108641. https://doi.org/10.1016/j.lwt.2019.108641

Morales-Contreras, B. E., Wicker, L., Rosas-Flores, W., Gallegos-Infante, J. A., Contreras-Esquivel, J. C., & Morales-Castro, J. (2024). Enzymatic de-esterification and depolymerization of apple pomace (Malus domestica var. Blanca de Asturias) pectin impacts lipid digestion in vitro. *Food Hydrocolloids, 146*, 109263. https://doi.org/10.1016/j.foodhyd.2023.109263

Osorio, L. L. D. R., Flórez-López, E., & Grande-Tovar, C. D. (2021). The potential of selected Agri-food loss and waste to contribute to a circular economy: Applications in the food, cosmetic and pharmaceutical industries. *Molecules, 26*(2), Article 2. https://doi.org/10.3390/molecules26020515

Otálora-Orrego, D., & Martin G., D. A. (2020). Técnicas emergentes de extracción de β-caroteno para la valorización de subproductos agroindustriales de la zanahoria (Daucus carota L.): Una revisión. *Informador Técnico, 85*(1). https://doi.org/10.23850/22565035.2857

Portillo-López, R., Morales-Contreras, B. E., Lozano-Guzmán, E., Basilio-Heredia, J., Muy-Rangel, M. D., Ochoa-Martínez, L. A., Rosas-Flores, W., & Morales-Castro, J. (2021). Vegetable oils as green solvents for carotenoid extraction from pumpkin (Cucurbita argyrosperma Huber) byproducts: Optimization of extraction parameters. *Journal of Food Science, 86*(7), 3122–3136. https://doi.org/10.1111/1750-3841.15815

Rodríguez, A. B. B., Fuertes, M. M. P., Ramírez, G. E. M., Rodríguez, A. B. B., Fuertes, M. M. P., & Ramírez, G. E. M. (2023). Uso potencial de residuos agroindustriales como fuente de compuestos fenólicos con actividad biológica. *MediSur, 21*(6), 1322–1330.

Santos, E. E., Amaro, R. C., Bustamante, C. C. C., Guerra, M. H. A., Soares, L. C., & Froes, R. E. S. (2020). Extraction of pectin from agroindustrial residue with an ecofriendly solvent: Use of FTIR and chemometrics to differentiate pectins according to degree of methyl esterification. *Food Hydrocolloids, 107*, 105921. https://doi.org/10.1016/j.foodhyd.2020.105921

de Souza, M. A., Vilas-Boas, I. T., Leite-da-Silva, J. M., Abrahão, P. d. N., Teixeira-Costa, B. E., & Veiga-Junior, V. F. (2022). Polysaccharides in agro-industrial biomass residues. *Polysaccharides, 3*(1), Article 1. https://doi.org/10.3390/polysaccharides3010005

Teodoro, A. J. (2019). Bioactive compounds of food: Their role in the prevention and treatment of diseases. *Oxidative Medicine and Cellular Longevity, 2019*(1), 3765986. https://doi.org/10.1155/2019/3765986

Viñas-Ospino, A., López-Malo, D., Esteve, M. J., Frígola, A., & Blesa, J. (2023). Green solvents: Emerging alternatives for carotenoid extraction from fruit and vegetable by-products. *Food, 12*(4), Article 4. https://doi.org/10.3390/foods12040863

Yepes, S. M., Naranjo, L. J. M., & Sánchez, F. O. (2008). Valorización de residuos agroindustriales – frutas – en Medellín y el Sur del Valle del Aburrá, Colombia. 61.

Chapter 5
Treatment of Slaughterhouse Wastewater for Reuse and Waste Utilization as a Circular Economy Alternative

Diana Laura Robles Morales, María del Rocío Ramírez Vargas, and Alejandro Reyes Cervantes

Abstract In Mexico, a total of 838 slaughterhouses have been officially recorded, collectively producing large volumes of wastewater resulting from the slaughter and processing of an average of 20,771 animals per month across bovine, porcine, ovine, caprine, equine, and poultry species. This wastewater is characterized by high pollutant loads, including chemical oxygen demand (COD), total suspended solids (TSS), fats, and nutrients such as phosphorus and nitrogen. The direct discharge of untreated slaughterhouse wastewater (SWW) into surface water bodies has led to eutrophication, ecosystem degradation, and serious public health risks due to the proliferation of pathogenic microorganisms. To address this issue, biological treatment processes have gained prominence, leveraging the metabolic activity of microorganisms to break down organic pollutants while generating biogas as a form of bioenergy. It is estimated that slaughterhouse waste can yield up to 8.654 million m^3 of biogas annually, reducing both carbon footprint and energy demand. Furthermore, non-hazardous organic byproducts such as fats and proteins can be converted into nitrogen-rich organic fertilizers, enhancing agricultural productivity through improved nutrient availability and pesticide degradation. This chapter focuses on biological treatment strategies for SWW, emphasizing sustainable, circular-economy frameworks through the integration of process-treatment-byproduct-application pathways.

Keywords Circular economy · Reuse and waste utilization · Slaughterhouse wastewater

D. L. Robles Morales · A. Reyes Cervantes (✉)
Laboratorio Aprovechamiento Integral de Recursos Bióticos, Universidad Politécnica de Pachuca, Zempoala, Mexico
e-mail: alejandro.r@colegiocondederomero.edu.mx

M. del Rocío Ramírez Vargas
Dirección de Investigación, Innovación y Posgrado, Universidad Politécnica de Pachuca, Zempoala, Mexico

S. G. Ceballos Pérez et al. (eds.), *Sustainability in Latin America*,
https://doi.org/10.1007/978-3-032-09787-3_5

5.1 Introduction

Global meat production in 2023 reached 357 million tons, with this industry being one of the most environmentally impactful. Water pollution from this industry generates effluents characterized by high concentrations of organic matter (proteins, fats, carbohydrates, solids, among others) (Bustillo-Lecompte & Mehrvar, 2017). These pollutants when discharged into the sewer system without any prior treatment generate various ecological problems for flora and fauna, and they reduce soil fertility for the production of commercially and nutritionally significant crops, as well as causing eutrophication in water bodies, thereby affecting aquatic life (Gaikar et al., 2010). Currently, the treatment of effluents generated in slaughterhouses is carried out biologically through aerobic and anaerobic processes, which can be used as strains or consortia depending on the characteristics of the slaughterhouse wastewater. However, in biological treatments, the removal rates of contaminants directly depend on the activity of the microorganisms used (Bustillo-Lecompte & Mehrvar, 2017; Robles-Morales et al., 2021). Studies have reported up to 90% removal of total organic matter in the biological treatment of slaughterhouse wastewater (Cao & Mehrvar, 2011). However, in these studies, slaughterhouse wastewater has been previously diluted before treatment because high concentrations of organic matter led to strong inhibition of biological sludge, reducing process efficiency and increasing contaminant removal times (Del Pozo et al., 2003; Rodríguez-Méndez et al., 2017; Freitas et al., 2017; Liu et al., 2018). Microbial consortia have been utilized as biological treatments, possessing the capability to perform various functions that an individual population could not accomplish due to the diversity of species and synergistic interactions developed among microorganisms (Harrison et al., 2006; Zhang et al., 2013). An example of such is the fungi/bacterial consortia, which exhibit high metabolic activity due to their clustering into granules, showing greater biomass retention without support media, allowing them to thrive in unfavorable environmental conditions. Additionally, they achieve higher biomass concentration, better solid-effluent separation, substrate transformation, product elaboration, and waste elimination (Bashiri et al., 2018). Once the water treatment process is completed, the use and reuse of bacterial and fungal-derived products and biomass can lead to positive behaviors and impacts reducing pollution, climate change and adverse environmental effects, thus enhancing economic performance within the circular economy (CE) framework. Concluding, the treatment of these wastewater represents a valuable source for the circular economy in the agricultural realm, achieving water reuse suitable for agriculture, increasing soil productivity through the use of biomass rich in bacteria that promote plant growth or as biofertilizers. A symbiosis between the meat industry and agriculture is fostered, thus propelling a more efficient and sustainable circular economy.

5.2 Water Pollution

Water is an indispensable natural resource for daily life activities, with only 1% of the water on the planet suitable for consumption and the needs of living organisms. However, it's not just about the quantity, but also about its quality (Ponsadailakshmi et al., 2018). This becomes an attribute that measures the level of contamination in terms of its physical, chemical, and biological properties. Therefore, specific quality standards are necessary to determine its potential uses, whether for drinking water, recreation, agriculture, industry, or animal slaughter. The latter is characterized by consuming large amounts of water, with an estimated use of 1000 liters per animal slaughtered (Signori et al., 2007; García et al., 2009; Rowles et al., 2018). Frequently, animal slaughter sites discharge their wastewater into the sewer system because the effects on the local flora, fauna, and microbiota are unknown. In Mexico, there are officially 2872 wastewater treatment plants, of which around 20% are dedicated to the treatment of slaughterhouse wastewater, while the rest of these discharge their wastewater untreated (SENASICA, 2022; CONAGUA, 2022).

5.3 Slaughterhouse Wastewater

Slaughterhouse wastewater is generated by the process of animal slaughter, and its composition depends on the species being processed. However, generally this type of water is characterized by high concentrations of proteins, fats, oils, excrement, and rumen content, among other contaminants. They present high concentrations of COD in a range of 5–26.290 gO_2/L, alkaline pH, and intense reddish colors due to the high concentration of blood. The physicochemical characteristics of slaughterhouse wastewater reported are summarized in Table 5.1 (Robles-Morales et al., 2021), highlighting the main parameters evaluated in the wastewater. It is observed that the DBO_5 concentration (2000 mg/L) is very high compared to the composition of domestic wastewater mentioned in the Official Mexican Norm NOM-002-Ecol-1996 for sewer discharge, which should be below 400 mg/L. One of the most important characteristics of wastewater is biodegradability, which can be established with the BOD_5/COD ratio. Municipal wastewater typically has BOD_5/COD values of 0.6–0.9, while the ratio in slaughterhouse wastewater ranges from 0.3 to 0.5, indicating low biodegradability. This is due to the high concentration of blood, which slows down the process due to the biological oxidation of organic matter

Table 5.1 Parameters of wastewater samples from Mineral del Monte, Hidalgo (Robles-Morales et al., 2021)

Parameter	Period		Mean
	Sacrifice	Maintaining	
COD (gL^{-1})	26.29 ± 1.66	2.80 ± 3.82	14.54
Protein (gL^{-1})	8.87 ± 1.88	1.175 ± 2.05	5.02
Carbohydrate (gL^{-1})	2.62 ± 0.15	0.325 ± 2.74	1.47
Fat (gL^{-1})	1.15 ± 0.13	–	0.57
pH	6.5 ± 0.01	7.2 ± 0.2	6.85

(blood), leading to the release of ammonia, which can alter the pH and inhibit the methanogenesis process (Del Pozo et al., 2003; Raffo & Ruiz, 2014). In our country, the volumetric flow discharged into the drains during the animal slaughter process is 15,010,586 L/day nationwide. Additionally, the pollutant load is 23,452.85 kg/day. This indicates a serious environmental issue in Mexico that needs to be studied and resolved, as slaughterhouses have a high demand for fresh water, representing competition for the available water supply for the population.

5.4 Harmful Effects of Slaughterhouse Wastewater

Slaughterhouse wastewater (SWW) generates various harmful effects, not only on overall health but also on microbiota and the environment. On the other hand, microorganisms present in these effluents are affected by high concentrations of proteins, which cause inhibition of methanogenic activity due to protein catabolism where ammonia (NH4+) is generated. Ammonia inhibits the methane synthesis enzyme and can passively diffuse into the cell, causing proton imbalance and potassium deficiency. This type of inhibition is associated with diffusion capacity in microbial cells and alteration of cellular homeostasis (Cuetos et al., 2008). Additionally, long-chain fatty acids (LCFA) can induce toxicity in anaerobic biomass. This process is somewhat similar to the toxicity caused by surfactants, leading to cell lysis and impairing sludge activity through LCFA absorption/adsorption processes, thereby limiting nutrient transfer across the cell membrane (Palatsi et al., 2009).

Furthermore, there is the formation of leachates that carry contaminants toward surface water bodies, drains, and sewers. Similarly, they can infiltrate into aquifers, deteriorating water supply sources for human consumption and irrigation of agricultural fields. Additionally, it causes eutrophication and low levels of available oxygen in the water, resulting in the death of aquatic flora and fauna (Signori et al., 2007; Bustillo-Lecompte & Mehrvar, 2017). These actions accelerate climate change, as primarily in the organic matter transformation processes greenhouse gases such as methane and carbon dioxide are emitted. This contributes to ozone depletion, acid rain, foul odors, and health risks (Rahman et al., 2014). Hence, there is a need to treat wastewater to reduce the effects it causes.

5.5 Biological Treatments of Slaughterhouse Wastewater (SWW)

Aerobic and anaerobic biological processes are an option in the treatment of slaughterhouse wastewater due to the economic advantages they represent. Additionally, they can be carried out using axenic cultures or mixed cultures. Different types of microorganisms have been reported for the removal of contaminants. These

Table 5.2 Microorganisms and consortia used in the treatment of slaughterhouse wastewater

Microorganisms	Type of process	Observations	References
Microalgae	Photobioreactors	Removal efficiencies of 91.7% with an initial concentration of 1.62 gL^{-1} of TOC and a HRT of 10 days	Hernández et al. (2016)
Thraustochytrium Kinney	Batch reactors	Removal efficiencies of 56.29% with an initial concentration of 9.61 gL^{-1} of COD and a HRT of 7 days	Villarroel and Silva (2018)
Anaerobic sludge	Biorreactors of 500 mL	Removal efficiencies of 90% with an initial concentration of 1.49 gL^{-1} of COD and a HTR of 30 days	Vidal et al. (2016)
Aerobic granular sludge	Bubble column reactor	Removal efficiencies of 89.9% with an initial concentration of 1.25 gL^{-1} of COD in 120 days	Liu et al. (2015)
Aerobic granules	Sequential batch reactor (SBR)	73%, with an initial concentration of 1.27 gL^{-1} of COD in 62 days	Bashiri et al. (2018)
Aerobic and anaerobic sludges	Anaerobic amphibious reactor and aerobic reactor	Removal efficiencies of 85.03%, with an initial concentration of 0.34 gL^{-1} COD in 75 days	Bustillo-Lecompte and Mehrvar (2017)
Carbonaceous nitrifying and denitrifying bacteria, aerobic and anaerobic	Sequential aerobic-anoxic reactor	Removal efficiencies of 86–95%, with an initial concentration of 0.087 gL^{-1} de NH_4^+ in 5 hours	Kundu et al. (2013)
Clostridium spp., *Parabacteroides* spp., *Desulfobulbus* spp., *Desulfomicrobium* spp., *Desulfovibrio* ssp., and *Desulfotomaculum* ssp.	Upflow anaerobic filter	Not described	Jabari et al. (2016)
Lichtheimia corymbifera, Cladosporium cladosporioides, Cladosporium tenuissimum, aspergillus versicolor and *Geotrichum candidum, Lysinibacillus fusiformis*	Erlenmeyer flasks (250 mL)	92% in 250 hours	Robles-Morales et al. (2021)

treatments have utilized bacteria, fungi, bacterial and fungal consortia, and activated sludge. These are described in Table 5.2.

The use of these consortia offers various advantages, since they perform different complex roles that an individual population could not carry out. One of them is its association, because it can generate greater resistance to environmental fluctuations and promote stability among community members. Likewise, they have the ability to withstand nutrient limitation periods due to the metabolic diversity available from the variety of species (Maier, 2009; Salgot & Folch, 2018). These consortia, which are typically built through a "bottom-up" approach, cultivate two or more isolated and metabolically characterized strains. Furthermore, they can be constructed through a "top-down" approach which involves enriching a stable microbial community, where each member is isolated and characterized afterward (Fig. 5.1).

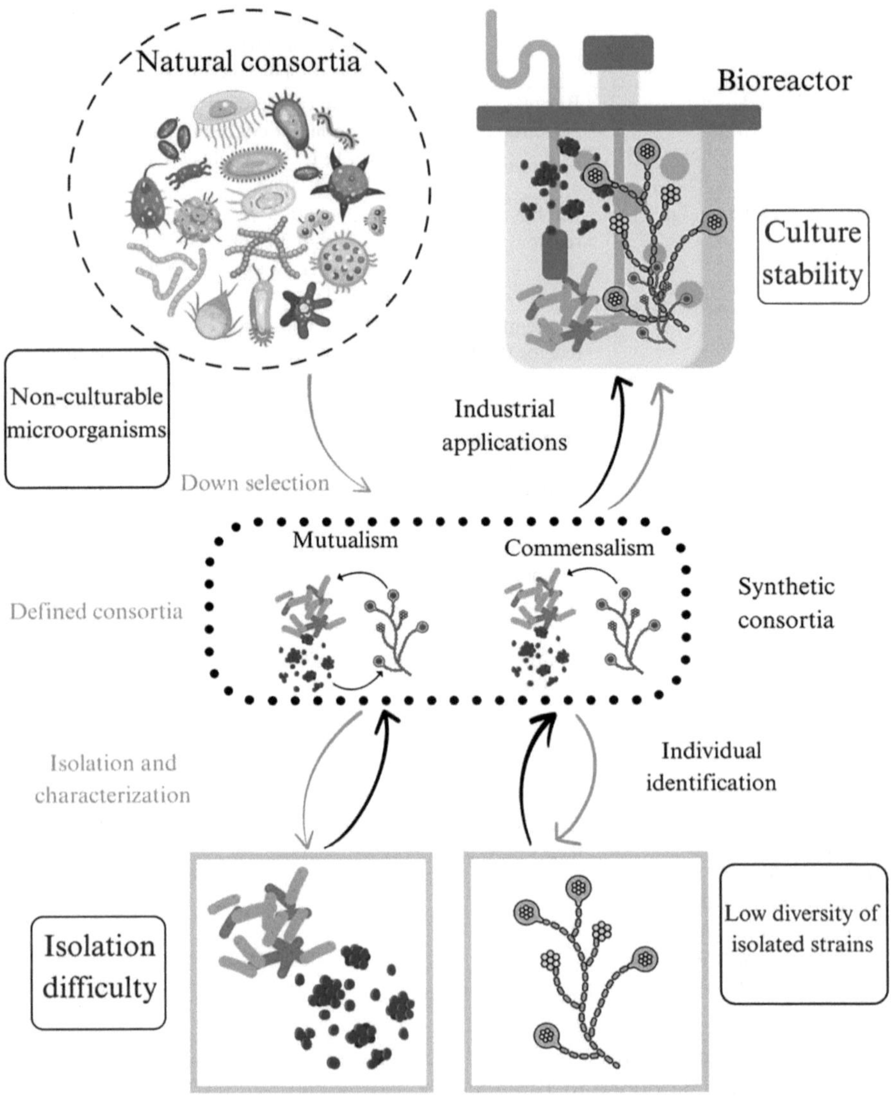

Fig. 5.1 Comparison of the "top-down" and "bottom-up" approaches for building microbial consortia. Strategies for formulating bioprocesses with the aim of providing a good or service and generating products that have an impact on the circular economy of the place or process where they are implemented

Through this process, biomass, byproducts, and metabolites are generated which can have another purpose.

5.6 Valorization of Slaughterhouse Wastewater Treatment to Stimulate the Circular Economy

In this section, opportunities for reusing processed water from slaughterhouses and converting the biomass generated into materials or applications of added value are discussed, which will help us to promote reuse, recycling, and reduce pollution. The research in this context focused on finding new applications for the advancement in the circular economy of biomass and wastewater treatment. In order to use slaughterhouse wastewater in efficient processes that provide benefits, it is important to look toward the field of biofuels. This is due to the fact that fossil fuel resources are limited and combustion is a significant source of environmental pollution. Therefore, a large part of the studies involving substrates rich in organic matter are an alternative for the generation of clean and sustainable energies. Among these substrates, commonly referred to as waste, we find those that come from the raising and slaughtering of animals. In these studies, it has been analyzed that slaughterhouse waste contains rumen, intestines, stomach, and blood of livestock. As a result, the production of biogas is feasible, since biogas production could reach up to 9 million m^3. This could be achieved from the 54 million m^3 of waste produced annually in these industries (Afazeli et al., 2014).

On the other hand, within the consortium used in the research projects of our working group, we find strains such as *Lysinibacillus fusiformis*. This microorganism has shown the ability to be efficient in the degradation of compounds such as organochlorine herbicides, specifically diuron, achieving removal efficiencies of this carcinogenic compound exceeding 90% using 360 mgL^{-1} of contaminant (Reyes-Cervantes et al., 2021).

On the other hand, the biomass obtained in these consortia shows the presence of various species of microorganisms such as *Cladosporium*, *Bacillus*, or *Aspergillus*, which have the ability to be used in efficient processes for the elimination of contaminants or producers of biotechnologically relevant metabolites of high added value (Table 5.3).

Another application of the byproducts obtained at the end of slaughterhouse wastewater treatment is their use in crop fields by adding the treated water. This water contains low concentrations of organic matter, rich in nutrients such as nitrogen and carbon, as well as microorganisms that can be used as growth promoters, given the presence of *Bacillus* species, which have the ability to solubilize nutrients, act as biological control, and promote plant growth (Robles-Morales et al., 2021; Ortiz & Sansinenea, 2024).

This type of water can be used in the agricultural sector. Taylor et al. (2018) demonstrated that treated water from a brewery industry was efficiently used in the

Table 5.3 Bioproducts or applications of the microorganisms used in microbial consortia

Microorganism	Bioproducts or application	Suggested use	Reference
Cladosporium	Compounds derived from mevalonic acid	Inhibition of *E. coli* growth activity	Zhang et al. (2023a, b)
Cladosporium sp.	Polyester degradation	Plastic degradation	Liu et al. (2023)
Aerobic and anaerobic sludges	Organic matter removal and water recovery	Wastewater treatment and removal of organic matter	Montalvo et al. (2018)
Aerobic and anaerobic sludges	Biogas	Industry	Zhang et al. (2023a, b)
Lysinibacillus	Polyethylene and poly-propylene elimination	Environment	Jeon et al. (2021)
Lysinibacillus	Diuron elimination	Environment	Reyes-Cervantes et al. (2021)
Aspergillus	Quinazoline alkaloids	Antimicrobial	Liao et al. (2024)
Microbial consortia	Polycyclic aromatic hydrocarbons	Environment	Zhou et al. (2024)

irrigation and growth of cabbage. Similarly, Kolekar et al. (2023) found that when using treated water, the growth of various crops was better than when using conventional water, as it was found that the nutritional values of the plants were better compared to the control group, containing proteins, Ca, Mg, Na, K, and P and improves tomato production by 59%. On the other hand, the reuse of wastewater for irrigation becomes necessary due to the scarcity of fresh water and depletion of underground resources.

The use of this type of wastewater can be an alternative to mitigate scarcity processes and has various advantages for farmers' profits and water resource preservation (Mishra et al., 2023). This type of wastewater has great agroeconomic potential regarding its use in the agrobiotechnological field. It presents benefits such as nutrient recovery, which are essential for the growth of plant species, as it has been shown to promote crop growth, improve soil fertility, and offer the possibility of obtaining quality food without dependence on chemical fertilizers. Additionally, it reduces the presence of contaminants and the elimination of pathogenic microorganisms, making the treated water safe for irrigation use. Furthermore, it promotes the reuse of water resources by obtaining microbial biomass, which can be used as bio-stimulants in the soil and act as biocontrol agents for pests and diseases.

5.7 Conclusions

The increasing demand for meat production has generated high amounts of slaughterhouse wastewater, increasing the water footprint and presenting an opportunity for treatment to encourage a circular economy. This type of wastewater offers interesting options for reuse after treatment or generating microorganisms that can be applied in other bioprocesses. It is possible to incorporate different processes using this type of wastewater as raw material, meaning the organic components of the water, including N and P, are utilized. If discarded untreated, these components can promote eutrophication of water bodies; however, their reuse could serve as a growth source for algae and/or plants, providing high-value metabolites like proteins, pigments, biofuels, and for formulating N-rich biofertilizers for crops such as tomatoes and corn. Additionally, microbial biomass is generated, which could be used as plant growth-promoting microorganisms and can also double as agents capable of metabolically removing contaminants from soils and water bodies (Fig. 5.2). These applications have been evaluated, showing contaminant removal rates of around 90%. Therefore, the biotransformation and recycling of this resource could help promote a circular economy, contributing to the reduction of pollutants, the reuse of water in irrigation systems, and the recycling of contaminant to obtain new value-added products.

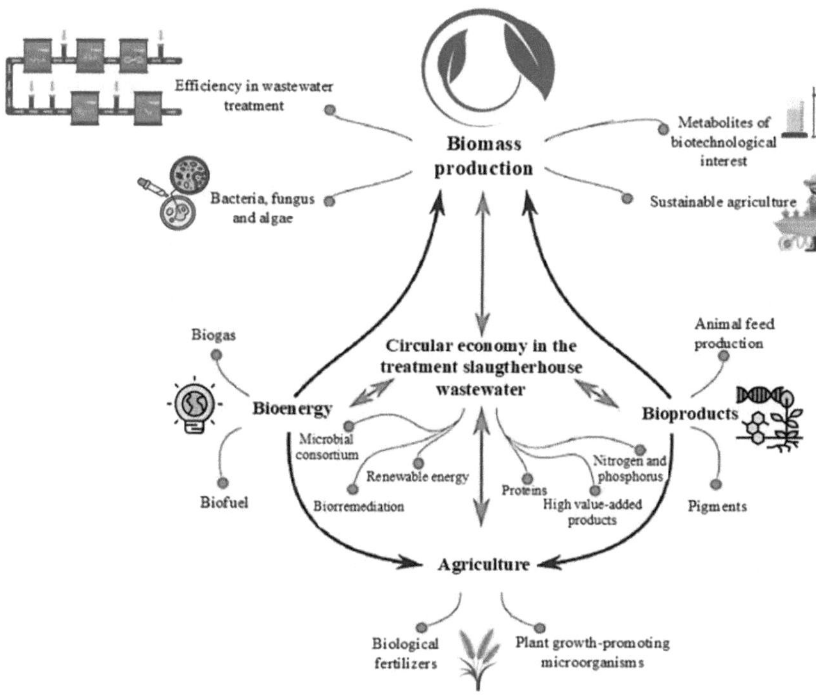

Fig. 5.2 Circular economy in the treatment slaughterhouse wastewater

References

Afazeli, H., Jafari, A., Rafiee, S., & Nosrati, M. (2014). An investigation of biogas production potential from livestock and slaughterhouse wastes. *Renew Sustain Energy, 34*, 380–386.

Bashiri, B., Fallah, N., Bonakdarpour, B., & Elyasi, S. (2018). The development of aerobic granules from slaughterhouse wastewater in treating real dyeing wastewater by Sequencing Batch Reactor (SBR). *Journal of Environmental Chemical Engineering, 6*(4), 5536–5543. https://doi.org/10.1016/j.jece.2018.05.020

Bustillo-Lecompte, C. F., & Mehrvar, M. (2017). Treatment of actual slaughterhouse wastewater by combined anaerobic–aerobic processes for biogas generation and removal of organics and nutrients: An optimization study towards a cleaner production in the meat processing industry. *Journal of Cleaner Production, 141*, 278–289. https://doi.org/10.1016/j.jclepro.2016.09.060

Cao, W., & Mehrvar, M. (2011). Slaughterhouse wastewater treatment by combined anaerobic baffled reactor and UV/H2O2 processes. *Chemical Engineering Research and Design, 89*(7), 1136–1143. https://doi.org/10.1016/j.cherd.2010.12.001

CONAGUA. (2022). Inventario Nacional de Plantas Municipales de Potabilización y de Tratamiento de Aguas Residuales en Operación. Available at http://www.chrome-extension://efaidnbmnnnibpcajpcglclefindmkaj/https://files.conagua.gob.mx/conagua/publicaciones/Publicaciones/SGAPDS-8-23.pdf

Cuetos, M. J., Gómez, X., Otero, M., & Morán, A. (2008). Anaerobic digestion of solid slaughterhouse waste (SHW) at laboratory scale: Influence of co-digestion with the organic fraction of municipal solid waste (OFMSW). *Biochemical Engineering Journal, 40*(1), 99–106. https://doi.org/10.1016/j.bej.2007.11.019

Del Pozo, R., Taş, D. O., Dulkadiroğlu, H., Orhon, D., & Diez, V. (2003). Biodegradability of slaughterhouse wastewater with high blood content under anaerobic and aerobic conditions. *Journal of Chemical Technology and Biotechnology, 78*(4), 384–391. https://doi.org/10.1002/jctb.753

Freitas, D. C., Passig, F. H., Kreutz, C., Carvalho, K. Q., Arantes, E. J., & Gomes, S. D. (2017). Effect of hydraulic retention time on hydrodynamic behavior of anaerobic-aerobic fixed bed reactor treating cattle slaughterhouse effluent. *Acta Scientiarum Technology, 39*(4), 469. https://doi.org/10.4025/actascitechnol.v39i4.30995

Gaikar, B., Uphade, B. K., Gadhave, A. G., & Kuchekar, R. (2010). Effect of dairy effluent on seed germination, seedling growth and biochemical parameter in paddy. *Rasayan Journal of Chemistry, 3*(1), 137–139.

García, J., Hinojosa-Huerta, O., Mclntyre, C., Zamora, F., Cheong, H., Shenahan, S., Zhou, X., Ramírez, J., Glenn, E., & Laura-Lopez, H. (2009). Ciénega de Santa Clara: Programa Integral de Monitoreo. 13–15. Available at http://centro.paot.org.mx/documentos/semarnat/cienega.pdf

Harrison, J. J., Turner, R. J., M. L. L., & C. H. (2006). Biopelículas. *Revista De Investigación Y Ciencia, 354*, 76–83.

Hernández, D., Riaño, B., Coca, M., Solana, M., Bertucco, A., & García-González, M. C. (2016). Microalgae cultivation in high rate algal ponds using slaughterhouse wastewater for biofuel applications. *Chemical Engineering Journal, 285*, 449–458. https://doi.org/10.1016/j.cej.2015.09.072

Jabari, L., Gannoun, H., Khelifi, E., Cayol, J. L., Godon, J. J., Hamdi, M., & Fardeaub, M. L. (2016). Bacterial ecology of abattoir wastewater treated by an anaerobic digestor. *Brazilian Journal of Microbiology, 47*(1), 73–84. https://doi.org/10.1016/j.bjm.2015.11.029

Jeon, J., M., Park, S. J., Choi, T. R., Park, J. H., Yang, Y. H., & Yoon, J. J. (2021). Biodegradation of polyethylene and polypropylene by *Lysinibacillus* species JJY0216 isolated from soil grove. *Polymer Degradation and Stability, 191*, 109662. https://doi.org/10.1016/j.polymdegradstab.2021.109662

Kolekar, S., Sankapal, P., Khare, K., & Chinnasamy, P. (2023). Assessment of recycled treated wastewater for sustainable tomato crop production: A comprehensive review. *Environmental Challenges, 13*, 100791. https://doi.org/10.1016/j.envc.2023.100791

Kundu, P., Debsarkar, A., & Mukherjee, S. (2013). Treatment of slaughter house wastewater in a sequencing batch reactor: Performance evaluation and biodegradation kinetics. *BioMed Research International, 2013*, 134872. https://doi.org/10.1155/2013/134872

Liao, X. L., Huang, J. G., Liu, P. Q., Yao, M., Wang, W. J., & Yang, X. L. (2024). Two new quinazoline alkaloids produced by *Aspergillus versicolor* and their antimicrobial activities. *Journal of Asian Natural Products Research, 26*(3), 320–327. https://doi.org/10.1080/10286020.2023.2230895

Liu, J., Zeng, Q., Lei, H., Xin, K., Xu, A., Wei, R., Li, D., Zhou, J., Dong, W., & Jiang, M. (2023). Biodegradation of polyester polyurethane by *Cladosporium* sp. P7: Evaluating its degradation capacity and metabolic pathways. *Journal of Hazardous Materials, 448*, 130776. https://doi.org/10.1016/j.jhazmat.2023.130776

Liu, G., Wang, J., & Campbell, K. (2018). Formation of filamentous microorganisms impedes oxygen transfer and decreases aeration efficiency for wastewater treatment. *Journal of Cleaner Production, 189*, 502–509. https://doi.org/10.1016/j.jclepro.2018.04.125

Liu, Y., Kang, X., Li, X., & Yuan, Y. (2015). Performance of aerobic granular sludge in a sequencing batch bioreactor for slaughterhouse wastewater treatment. *Bioresource Technology, 190*, 487–491. https://doi.org/10.1016/j.biortech.2015.03.008

Maier, R. M. (2009). Biogeochemical cycling. Chapter 14. In *Environmental microbiology* (pp. 287–318). https://doi.org/10.1016/B978-0-12-370519-8.00014-6

Mishra, S., Kumar, R., & Kumar, M. (2023). Use of treated sewage or wastewater as an irrigation water for agricultural purposes – Environmental, health, and economic impacts. *Total Environment Research Themes, 6*, 100051. https://doi.org/10.1016/j.totert.2023.100051

Montalvo, S., Vielma, S., Borja, R., Huiliñir, C., & Guerrero, L. (2018). Increase in biogas production in anaerobic sludge digestion by combining aerobic hydrolysis and addition of metallic wastes. *Renewable Energy, 123*, 541–548. https://doi.org/10.1016/j.renene.2018.02.004

Ortiz, A., & Sansinenea, E. (2024). 5 – *Bacillus* sp. as biofertilizers applied in horticultural crops. *Bio-Inoculants in Horticultural Crops, 3*, 97–108. https://doi.org/10.1016/B978-0-323-96005-2.00007-6

Palatsi, J., Laureni, M., Andrés, M. V., Flotats, X., Nielsen, H. B., & Angelidaki, I. (2009). Strategies for recovering inhibition caused by long chain fatty acids on anaerobic thermophilic biogas reactors. *Bioresource Technology, 100*(20), 4588–4596. https://doi.org/10.1016/j.biortech.2009.04.046

Ponsadailakshmi, S., Sankari, S. G., Prasanna, S. M., & Madhurambal, G. (2018). Evaluation of water quality suitability for drinking using drinking water quality index in Nagapattinam district, Tamil Nadu in Southern India. *Groundwater for Sustainable Development, 6*, 43–49. https://doi.org/10.1016/j.gsd.2017.10.005

Raffo, L. E., & Ruiz, L. E. (2014). Caracterización de las aguas residuales y la demanda bioquímica de oxígeno. *Revista De La Facultad De Ingenieria Industrial, 17*, 11. Available at http://www.redalyc.org/pdf/816/81640855010.pdf

Rahman, U., Sahar, A., & Khan, M. A. (2014). Recovery and utilization of effluents from meat processing industries. *Foodservice Research International, 65*(Part C), 322–328. https://doi.org/10.1016/j.foodres.2014.09.026

Reyes-Cervantes, A., Robles-Morales, D. L., Téllez-Jurado, A., Huerta-Ochoa, S., Jiménez-González, A., & Medina-Moreno, S. A. (2021). Evaluation in the performance of the biodegradation of herbicide diuron to high concentrations by *Lysinibacillus fusiformis* acclimatized by sequential batch culture. *Journal of Environmental Management, 291*, 112688. https://doi.org/10.1016/j.jenvman.2021.112688

Robles-Morales, D. L., Reyes-Cervantes, A., Díaz-Godínez, R., Medina-Moreno, S. A., & Jiménez-González, A. (2021). Design and performance evaluation of a fungi-bacteria consortium to biodegrade organic matter at high concentration on synthetic slaughterhouse wastewater. *Water, Air, and Soil Pollution, 232*(6). https://doi.org/10.1007/s11270-021-05177-1

Rodríguez-Méndez, R., Le Bihan, Y., Béline, F., & Lessard, P. (2017). Long chain fatty acids (LCFA) evolution for inhibition forecasting during anaerobic treatment of lipid-rich wastes: Case of milk-fed veal slaughterhouse waste. *Waste Management, 67*, 51–58. https://doi.org/10.1016/j.wasman.2017.05.028

Rowles, L. S., Alcalde, R., Bogolasky, F., Kum, S., Diaz-Arriaga, F. A., Ayres, C., Mikelonis, A. M., Toledo-Flores, L. J., Alonso-Gutiérrez, M. G., Pérez-Flores, M. E., Lawler, D. F., Ward, P. M., Lopez-Cruz, J. Y., & Saleh, N. B. (2018). Perceived versus actual water quality: Community studies in rural Oaxaca, Mexico. *Science of the Total Environment, 622–623*, 626–634. https://doi.org/10.1016/j.scitotenv.2017.11.309

Salgot, M., & Folch, M. (2018). Wastewater treatment and water reuse. *Current Opinion in Environmental Science & Health, 2*, 64–74. https://doi.org/10.1016/j.coesh.2018.03.005

SENASICA. (2022). Listado de rastros y/o centros de matanza en los que se realiza vigilancia o seguimiento operativo por parte de las entidades federativas. Available at https://www.gob.mx/cms/uploads/attachment/file/736767/LISTADO_RASTROS_P_BLICO_220622.pdf

Signori, M., Civit, S., Bonilla, M., Cervantes, M., Calderon, M., Perez, A., Espejel, M., Almanza, C., Porchietto, M. S., Gual, S. C., Padilla, M. B., Elena, M., Ramírez, C., Vázquez, M. C., Montecillo, A. P., & Rodríguez, C. A. (2007). De Los Rastros Y Mataderos Municipales. *NACAMEH, 1*(2), 118–141.

Taylor, R. P., Jones, C. L. W., Laing, M., & Dames, J. (2018). The potential use of treated brewery effluent as a water and nutrient source in irrigated crop production. *Water Resources and Industry, 19*, 47–60. https://doi.org/10.1016/j.wri.2018.02.001

Vidal, J., Huiliñir, C., & Salazar, R. (2016). Removal of organic matter contained in slaughterhouse wastewater using a combination of anaerobic digestion and solar photoelectro-Fenton processes. *Electrochimica Acta, 210*, 163–170. https://doi.org/10.1016/j.electacta.2016.05.064

Villarroel, H., M. P., & Silva, D. (2018). Bioremediation of piggery slaughterhouse wastewater using the marine protist, *Thraustochytrium Kinney* VAL-B1. *Journal of Advanced Research, 12*, 21–26. https://doi.org/10.1016/j.jare.2018.01.010

Zhang, D., Zheng, J., Chu, M., Feng, Y., & Li, H. (2023a). Secondary metabolites from the endophytic fungus *Cladosporium* sp. ZLM-6. *Biochemical Systematics and Ecology, 111*, 104741. https://doi.org/10.1016/j.bse.2023.104741

Zhang, X., Yang, H., Wie, D., Chen, Z., Wang, Q., Song, Y., Ma, Y., & Zhang, H. (2023b). Tolerance of anaerobic digestion sludge to heavy metals: COD removal, biogas production, and microbial variation. *Journal of Water Process Engineering, 55*, 104157. https://doi.org/10.1016/j.jwpe.2023.104157

Zhang, L., Hu, J., Zhu, R., Zhou, Q., & Chen, J. (2013). Degradation of paracetamol by pure bacterial cultures and their microbial consortium. *Applied Microbiology and Biotechnology, 97*(8), 3687–3698. https://doi.org/10.1007/s00253-012-4170-5

Chapter 6
Biocultural Relevance and Bioeconomic Perspectives of Ahuilote (*Vitex mollis* Kunth)

Blanca Catalina Ramírez Hernández, Sara Villanueva Viramontes, Ernesto Ramírez Briones, Paulina Beatriz Gutiérrez Martínez, and Javier Eugenio García de Alba Verduzco

Abstract The bioeconomy provides a platform for reflecting on the reassessment or valuation of local biocultural resources, such as the ahuilote (*Vitex mollis* Kunth). Despite being overlooked in certain areas, this fruit of pre-Hispanic heritage presents an opportunity in the Valles region of Jalisco, Mexico, for the utilization and integration of cultural knowledge with a sustainable approach to natural resource use. The significance of ahuilote is highlighted in Ahualulco de Mercado and Etzatlán, in the Valles region, where the "Ahuilote Fair" is held in the former municipality. A qualitative methodology was employed, involving semi-structured interviews with key informants, such as gatherers, vendors, and artisans, to gather insights into traditional and current uses of the fruit. Informants noted that the fruit has traditionally been used for fresh consumption, processing, and particularly for its medicinal properties (both the fruits and the leaves). Its aromatic leaves, attractive violet or blue flowers, and sweet fruits, which darken when ripe, have contributed to a long history of local use. However, this tradition may diminish due to a lack of recognition by new generations. This species can serve as a model to expand the understanding of the economy-nature connection in regions transitioning to a utilitarian model of natural resources. This chapter explores how biocultural heritage can foster a plural, sustainable, and environmentally responsible bioeconomic development, demonstrating how the local circular bioeconomy approach aligns environmental sustainability with social, cultural, and economic advancement.

B. C. Ramírez Hernández (✉) · S. Villanueva Viramontes · E. Ramírez Briones · J. E. García de Alba Verduzco
Department of Applied Ecology, University of Guadalajara, Zapopan, Mexico
e-mail: blanca.ramirez@academicos.udg.mx; sara.villanueva@academicos.udg.mx; ernesto.ramirez@academicos.udg.mx; javier.garciadealba@academicos.udg.mx

P. B. Gutiérrez Martínez
Department of Environmental Sciences, University of Guadalajara, Zapopan, Mexico
e-mail: paulina.gutierrez@academicos.udg.mx

S. G. Ceballos Pérez et al. (eds.), *Sustainability in Latin America*,
https://doi.org/10.1007/978-3-032-09787-3_6

Keywords Regional development · Sustainable · Food · Circular economy · Agrobiodiverse

6.1 Introduction

The Valles region of Jalisco serves as a fascinating study area within the state, where diverse biological and cultural expressions intersect. This includes the "Sierra del Águila" State Hydrological Protection Area, the RAMSAR site "Presa de la Vega," the archaeological zones of Guachimontones, and the "Ocomo" palace (Sánchez Moreno, 2015), alongside significant agricultural and livestock activities, particularly the cultivation of sugarcane and agave (Uvario et al., 2018). The former has been cultivated for centuries, characterized by traits that led to its inclusion in the World Heritage list in 2006 as the agave landscape and ancient industrial facilities of Tequila. In contrast to other regions of Mesoamerica, the western part of Mexico has not been explored as thoroughly. However, archaeological evidence suggests that pre-Hispanic populations inhabited Jalisco and other areas of western Mexico since 1500 B.C. (Tetreault & López, 2024). Findings from western Mexico indicate that indigenous groups lived in dispersed villages, predominantly organized into small-scale chiefdoms, with limited evidence of domestication and consumption of plant species. It is known that the "roza, tumba, quema" technique was commonly employed to produce temporary crops, especially maize, beans, and squash. Additionally, they collected honey, gathered wild fruits, practiced hunting, and crafted various artisanal items (Tetreault & López, 2024). With the arrival of the Spaniards in the Valles region at the end of the sixteenth century, indigenous groups experienced transformations due to various aspects of daily life, including conflicts among the different chiefdoms of the region, diseases, and trade exchanges with other areas of Mexico, such as with the P'urhépecha culture (Acuña, 2017). While this mixture brought various favorable conditions to the region, the geographical reports of Nueva Galicia emphasized that the original peoples of this area did not readily adopt or thrive with crops such as apples, figs, grapes, citrus, olives, and wheat, among others. Instead, it is noted that they continued to consume local foods and fruit trees, including varieties of plums, zapotes, guamuchiles, and mesquites. Such descriptions highlight the richness of plant-based foods consumed in the area, suggesting, according to the geographical reports of Nueva Galicia, that the native people lived up to 100 years. Although there is no clear reference to fruits like ahuilote, it is important to note that its traditional consumption holds significance in this region of Western Mexico. Furthermore, while there is no explicit mention of ahuilotes in the Natural History of New Spain or the Codex of the Cruz-Badiano, it is an ancestral fruit with traditional uses that are undoubtedly reflected through various oral traditions of the cultures along the Pacific slope of Mexico, from southern Sonora to Oaxaca. In the current state of Jalisco, the continuation of indigenous traditions is a complicated process filled with historical, ethnobiological, and cultural challenges. Over the past five centuries, the indigenous groups that inhabited the present territory of Jalisco underwent drastic transformations during three key

Fig. 6.1 Ahuilote fruits (*Vitex mollis*) from Ahualulco de Mercado and Etzatlán, Valles de Jalisco region

moments: the wars and diseases of the conquest, the colonial era policies that notably altered their diet, and finally, the post-revolutionary projects that, alongside land restitution, introduced "acculturation" and "integration" programs—led by the National Indigenous Institute (INI)—where the premise was "de-indianization," characterized by the loss of indigenous languages and identities. This significantly reduced the consumption of foods considered wild and that do not require substantial agronomic management (Tetreault & López, 2024; Zizumbo-Villarreal & Colunga-García Marín, 2010), among which we can include species like ahuilote. *Vitex mollis* is a tree that can reach up to 20 meters in height, producing dark purple fruits (Pennington & Sarukhán, 2005) that are nearly black (Fig. 6.1), with a strong aroma and flavor. It is known by various names in different regions of Mexico, such as capulincillo, nanche de perro (State of Mexico and Michoacán), atuto (Michoacán, Guerrero), uvalama or uvalamo (Nayarit, Sinaloa, and Sonora), coyotomate

(Morelos and Guerrero), beo (Oaxaca), and ahuilote (Jalisco) (Blancas et al., 2023). In the municipalities of the Valles region of Jalisco, Ahualulco de Mercado has been a pioneer in reviving the consumption and traditional practices associated with ahuilote, as its consumption has been on the decline. This phenomenon highlights a gap in knowledge and presents an opportunity to investigate and promote the significance of ahuilote in sustainable development and the conservation of the region's biocultural heritage, generating new circular economy strategies that encourage the efficient and sustainable use of resources (Zuniga Gonzalez & Jaramillo-Villanueva, 2024). However, to achieve this, it is essential to thoroughly understand the importance of ahuilote in local culture and economy, as well as the challenges and opportunities related to its utilization. The objective of this research was to explore the significance of ahuilote at Ahualulco de Mercado and Etzatlán in the Valles of Jalisco Region, focusing on promoting sustainable development, conserving biocultural heritage, and creating conditions for a circular bioeconomy. This was accomplished through a comprehensive literature review of studies on the fruit, along with interviews with key stakeholders and informants in municipalities that have a strong tradition of consumption. This approach offers a detailed and current perspective on the significance of ahuilote in the Valles region of Jalisco.

6.2 Material and Methods

Study Area. This research was conducted in the municipalities of Ahualulco de Mercado and Etzatlán, situated in the Valles region of Jalisco. This area features primary tree vegetation covering just over a quarter (26.68%) of the land, which remains in its natural state. The region consists of 12 municipalities. Visits were made to Etzatlán and Ahualulco de Mercado (Fig. 6.2) due to their records of fruit consumption, a growing interest in the species as a means to enhance regional development, and a significant number of documented wild *V. mollis* individuals (Martínez-Gordillo et al., 2017). Ahualulco de Mercado has a land cover that is predominantly agricultural (60.2%), followed by forests (19.77%) and secondary vegetation (17.42%). It contains two protected natural areas (5795.15 hectares; 24.6% of the territory). The municipal seat has a population of 17,000, with 33% living in poverty, 26.8% vulnerable due to social deficiencies, 13.9% vulnerable due to income, and 26.3% neither poor nor vulnerable (Instituto de Información Estadística y Geográfica de Jalisco, IIEG, 2022). In Etzatlán, secondary vegetation covers 46.46%, followed by forest at 21.96%, and human settlements occupy 0.68%. It has three protected natural areas (8590.46 hectares; 22.1% of its territory). The municipal seat has a population of 14,697, with 37.9% living in poverty, 34.6% vulnerable due to social deficiencies, 6.8% due to income, and 20.6% neither poor nor vulnerable (IIEG, 2023).

For this study, a participatory action research methodology was employed with an academia-community approach. This was accomplished through the active involvement of focus groups (Hernández Sampieri et al., 2010) or key informants from the

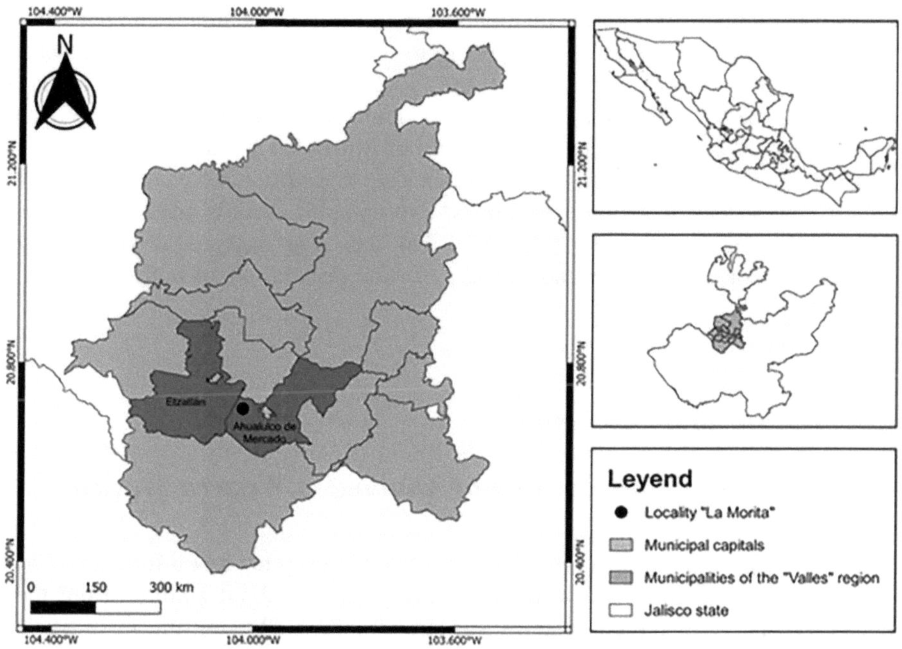

Fig. 6.2 Map of the Valles de Jalisco region

population centers of the municipalities of Etzatlán and Ahualulco de Mercado. This methodology facilitates the identification of issues related to the priority needs of the populations in areas such as food, health, appreciation of traditional knowledge (biological and ecological), income, and commercialization, among others, as it provides guidelines for interdisciplinary research teams to address these needs in the pursuit of local and regional solutions (Estrada-Martínez et al., 2020). To understand the social perception regarding the knowledge of ahuilote, its uses, management, conservation, commercialization, and perceived value, a semi-structured survey was designed to gather information from a sample of key informants (interviewees). These interviews were conducted with vendors and/or gatherers, as well as artisans who create value-added ahuilote products, and one of the organizers of the "Feria del Ahuilote" in the municipality of Ahualulco de Mercado. The survey was organized into three sections: (1) knowledge and use of the species; (2) commercialization; and (3) conservation and sustainability. Respondents were informed about the privacy policy and confidential data handling, ensuring the anonymity and confidentiality of their responses; this work was carried out under the framework of the Code of Ethics of the Latin American Society of Ethnobiology (Argueta et al., 2018).

The collected data relates to participants' perceptions, semi-structured interviews, and specific dialogues with gatherers, vendors from the municipal markets of both localities, and artisans who create and distribute value-added ahuilote products

(Taylor & Bogdan, 2002). This qualitative data collection facilitates an approximation of local knowledge through a perspective that shares the activity and space, conducted via direct contacts and the documentation of specific scenarios (Rodríguez, 2009). For this purpose, the commercialization of ahuilote fruits and/or products sold in the municipal markets of Ahualulco de Mercado and Etzatlán was documented. In addition to direct contacts, *V. mollis* trees were surveyed to identify the species in both communities. Photographic records and samples were gathered for future reference (Fig. 6.3). The data was analyzed from a gender perspective to gain a deeper understanding of the perceptions of both women and men.

6.3 Results and Discussion

6.3.1 Uses and Consumption of Ahuilote in Western México

For this study, voluntary participation was obtained from ten key informants. Most interviewees (both men and women) reported learning about ahuilote from their ancestors, specifically their mothers and grandmothers. Among the male participants, two mentioned discovering ahuilote through friends and media, while most women indicated they obtained ahuilote by collecting fallen fruits from wild trees; one man noted a preference for purchasing from other collectors. Notably, a craftsperson/collector from the municipality of Etzatlán not only collects but also buys ahuilotes to satisfy demand during the "Ahuilote Fair," held annually since 2022 and promoted by the municipality of Ahualulco de Mercado. She does this to pass down the knowledge she gained from her mother on how to select, collect, clean, and consume the fruit, enriched by her expertise in making preserves for consumers outside the locality who long for traditional flavors (D' Andrea & D'Ulizia, 2023).

Regarding the uses of this species, women have greater knowledge (Fig. 6.4). They reported medicinal uses ranging from leaf infusions for stomach aches or to induce sleep, roots for intestinal cleansing, to poultices with the bark to rehabilitate the abdomens of postpartum women. In terms of consumption frequency, women reported a higher frequency, being daily during the season (Fig. 6.4E). However, the way it is consumed is by cooking the fruits with piloncillo, cinnamon, and/or guava, whereas men mentioned more forms of consumption, from ripe fresh fruit, with lemon, salt, and chili, to cooked with piloncillo and mixed with milk (Fig. 6.4F). On the other hand, in terms of commercialization (Fig. 6.1), a larger number of women sell fresh ripe fruits in both municipal markets, with prices ranging from 30 to 40 pesos (MXN) (~ \$1.5 to \$2 USD) per liter of fresh fruits (Fig. 6.1C). However, the interviewed craftsperson from Etzatlán reported that three collectors sell the fruits door-to-door, not in the market. Both women and men sell fresh ripe fruits; however, women are the ones who sell value-added products, such as preserves with piloncillo, with sugar and cinnamon, with piloncillo and cinnamon, with stevia (a product made by a craftsperson from Etzatlán intended for consumption by

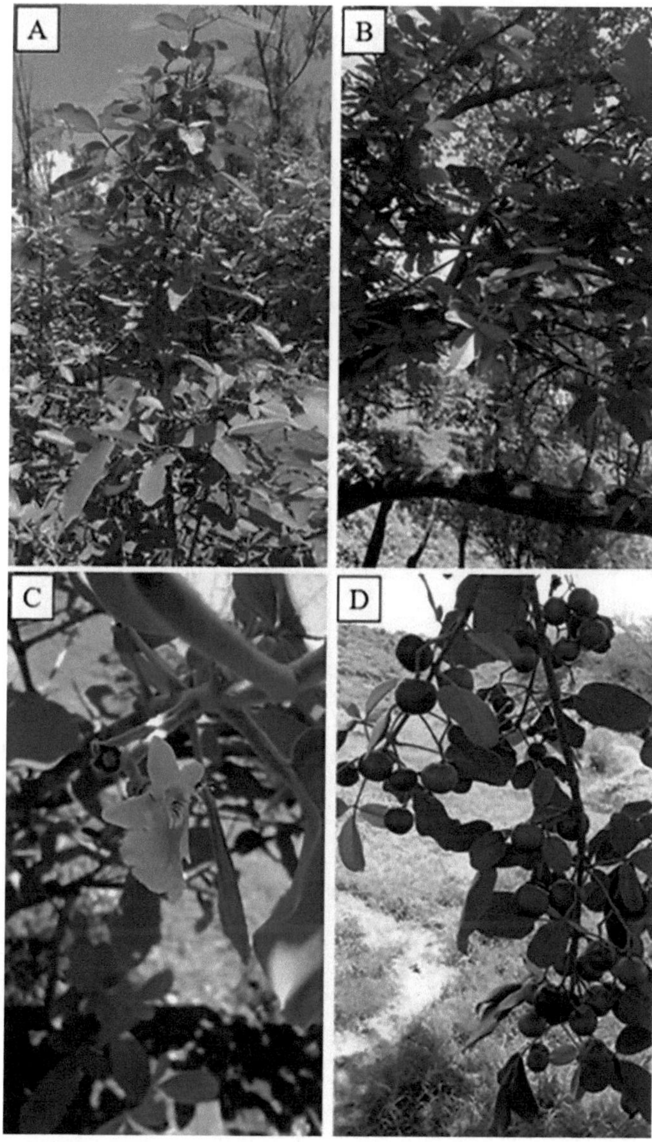

Fig. 6.3 Ahuilote tree, flowers and fruits (*Vitex mollis*)

diabetics), with guava (which the interviewees identify as the traditional preparation from Ahualulco de Mercado), crystallized pears with ahuilote sauce, ahuilote gelatin, milk gelatin with ahuilote cream, guava cream with ahuilote cream, and cheesecake with ahuilote sauce (priced up to 550 pesos (MXN) = ~ $27.5 USD). These latter products are made by young entrepreneurial women from Ahualulco de Mercado, aged between 20 and 30 years (not artisans), who were motivated by the "Ahuilote Fair" (information provided by the fair's manager) (D'Andrea & D'Ulizia, 2023).

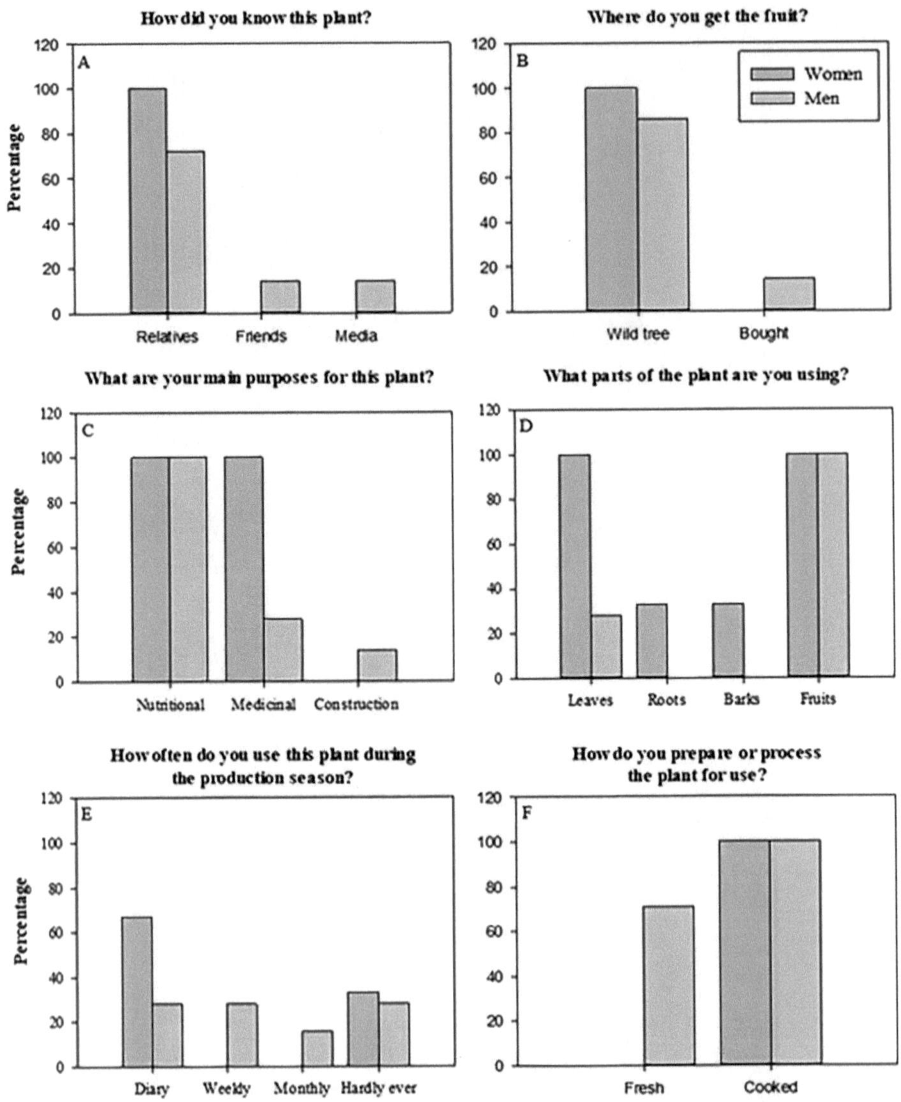

Fig. 6.4 Uses of ahuilote and frequency of consumption

Additionally, both men and women reported selling fresh ripe fruits for over 10 years, and a vendor from Etzatlán (approximately 70 years old) helped her mother sell the fruits in the market from an early age (10 years old). Regarding the perception of demand for ahuilote, both fresh ripe fruits and value-added products, women reported higher demand (from moderate to very high) compared to men's perception (from very low to high). It is worth mentioning that those women who reported a high to very high demand perception are directly involved in the Ahuilote

Fair and strive to promote its consumption due to its low acceptance among younger generations. It is important to highlight that all interviewees mention the connection between the consumption of ahuilote and their territory, as they are collected from the fields during the rainy season, taking advantage of family outings to the streams of municipalities such as Las Torcazas in Ahualulco de Mercado (Barros et al., 2021).

In line with Perales and Aguirre (2008), behind all the plant species that continue to be utilized by human groups in certain biodiversity-rich regions, as in our case study, there are interesting events that have promoted greater genetic diversity, as shown by archaeobotanical studies in regions distinguished by their biocultural wealth. Following the multiple historical events of western Mexico, the importance of the Valles region stands out, where significant pre-Hispanic settlements such as the Guachimontones and the palace of Ocomo flourished from the Early Classic period (300 B.C.–500 A.D.), which still require research, along with 31 other sites distributed in the center and south of a paleolake that extended across the current municipalities of Magdalena, Etzatlán, and San Juanito de Escobedo (Blanco & Pérez Martínez, 2022). During the Preclassic period in the Valles region (1500–300 B.C.), adaptation to the environment was key, and studies indicate an exploitation combining animal husbandry and the cultivation of plants and fruits likely supported by the waters of the extinct lake (Mountjoy, 2015). From the drying up of the lake during the nineteenth century (Domínguez, 1997), the landscape changed, leading to the birth of a new culturally recognized landscape, the agave landscape, which initially constituted a mix of traditional crops such as corn, beans, peanuts, fruit trees, and agave without the latter dominating, as it currently does (Orendain & Michel, 2018).

Without conclusive evidence regarding the consumption of ahuilote in western Mesoamerica since pre-Hispanic times, the information obtained in the municipalities of the Valles region, Jalisco, shows a tradition that persists and seeks revitalization through various strategies; particularly in the municipality of Ahualulco, where over the past 3 years, efforts have been coordinated between the municipal government and ahuilote marketers to create the "Ahuilote Fair" (an event that continues to grow). Here, the importance of transmitting these traditions to new generations is perceived (Barros et al., 2021), using cultural events to revitalize the local economy and strengthen the sense of community (Gillani et al., 2024) in various localities of Ahualulco de Mercado and Etzatlán, in line with the foundations of the bioeconomy that additionally aim to complement with the sustainable use of biological resources (Zuniga-González et al., 2022, 2024). In this context, the utilization of *V. mollis* plays an important role from an ecological, social, and bioeconomic perspective, as it is considered a sustainable form of exploitation. In an oligopsony-type agricultural environment, where previous initiatives aimed at energizing the local economy only managed to introduce a model based on greenhouse monocultures (Vides-Rodríguez et al., 2019), the emergence of alternative strategies centered on native species such as *V. mollis* represents an effort to diversity production and promote a more sustainable and inclusive local development model. On the other hand, native species offer an ecological, economic, and cultural space that allows traditional management and use in a more sustainable way. The use of the ahuilote is embedded in a regional development model coupled to the principles of bioeconomy. Nonetheless, it is essential to highlight the need for increased support to strengthen the economy of collectors, as well as to encourage research and technological development that promotes regional growth through the economic valuation of environmental goods and services (Rangel Cura et al., 2015).

6.3.2 Chemical Studies on Ahuilote, New Perspectives of Utilization

This tree, which provides multiple services, has been an integral part of the culture and diet of local communities since pre-Hispanic times in various regions of Mexico (Blancas et al., 2023), as well as in the Valles region of Jalisco according to the results of this work, becoming a key component for sustainable development and the conservation of the region's biocultural heritage. Ahuilote is a rich source of vitamins, minerals, and functional compounds, making it a nutritious and healthy food. Additionally, it contains compounds with anticancer, antibacterial, anti-inflammatory, and antiprotozoal activity, further enhancing its value and potential for human well-being (Delgado-Vargas et al., 2010; Montiel-Herrera et al., 2004).

For the genus *Vitex*, there are numerous medicinal properties mainly associated with fruits, which are often an important food resource in rural and low-income areas (Pio-León et al., 2014). These particular fruits from the Lamiaceae family provide vitamins, minerals, and fiber, in addition to various compounds that offer benefits in the prevention and treatment of certain diseases, making their consumption associated with a healthy fruit (Kaliora & Dedoussis, 2007; Morales-Del Rio et al., 2015). There are 70 species consumed worldwide, mainly in Africa, where their fruits are known as "Chocolate Berries"; these wild berries play an important role in food security, sustainable rural development, and the economy of various communities (National Research Council, 2008). Despite the large number of species in the genus *Vitex* (270 spp.), few have been studied in depth, such as *Vitex agnus-castus* L., which has information on compounds in fruits and by-products that are marketed in Europe and the United States to treat premenstrual syndrome. In the case of *V. mollis*, a species native to Mexico, chemical studies are limited and only partial characterizations of fruits have been conducted, which have identified the presence of flavonoids and melanins (Montes-Avila et al., 2018), tannins, and phytic acid in concentrations safe for human consumption (Montiel-Herrera et al., 2004). Such studies have been conducted on fresh fruits, which opens new opportunities since the main form of ahuilote consumption is in sweet preparations consumed alone or mixed with dairy products, and no chemical characterization exists yet (Montiel-Herrera et al., 2004). Proximal studies conducted on ahuilote revealed a significant content of minerals such as zinc and copper, as well as a high content of Vitamin C.

Contrasting *Vitex agnus-castus* and *Vitex rotundifolia* L., where flavonoids such as casticin and orientin, terpenes such as rotundifuran and ferrugin, iridoids such as aucubin, and essential oils such as sabinene have been identified, in *V. mollis* only the presence of phytic acid and melanins have been confirmed. This opens up possibilities for further studies that promote greater consumption as an important source of beneficial compounds through strategies that allow for the conservation of the species, more research, and the production of high-value-added by-products such as extracts and isolates (Pío-León et al., 2012).

6.3.3 Local Bioeconomy: Perspectives on Ahuilote Utilization

Our research demonstrates that ahuilote is an underutilized resource with considerable potential for the circular bioeconomy and the conservation of biocultural heritage in the Valles of Jalisco (Alexandri et al., 2023; Gomez San Juan et al., 2022). From a bioeconomy perspective, the inhabitants of Ahualulco de Mercado and Etzatlán have transformed the ahuilote into a key resource for generating community-based value, although they do not handle the concept of bioeconomy, beside the absence of a clear national policy, where a formal strategy linking sustainability with the utilization of biocultural resources has yet to be developed (De la Cruz Borrego & Caballero Rico, 2021).

In this context, beyond its traditional consumption, the fruit has given rise to a variety of processed products, such as preserves, jam, and beverages, which have strengthened local economic networks and enhanced its culinary significance. Unlike other extractivist models of natural resource utilization, these communities' relationship with the ahuilote is rooted in sustainability principles, ensuring that harvesting aligns with the tree's natural production cycles and prevents overexploitation. This approach integrates the use of ahuilote into broader circular economy strategies, promoting its conservation while providing direct economic benefits to harvesters and small-scale producers (Salvador et al., 2022).

This is the first comprehensive assessment of ahuilote, incorporating both a literature review and interviews with residents. The nutritional value of ahuilote, identified through proximal studies that revealed the presence of minerals such as copper and zinc, in addition to functional compounds like melanins (Montiel-Herrera et al., 2004), allows for exploitation routes in fields such as medicine, biotechnology, agriculture, and cosmetics, without leading to scenarios that may increase social and economic inequalities (Fundación Biodiversidad, 2024).

These results align with previous research that has highlighted the nutritional value of unconventional fruits and their potential to contribute to food security and human health, integrating science and indigenous and local knowledge. The aim is to promote bioeconomy through traditional conservation and management practices that help preserve biodiversity and ensure the fair distribution of benefits (Bracco et al., 2019), as established by the Nagoya Protocol (Meza & Rodríguez, 2022). Despite hypothesizing a proposal that uses ahuilote as a source of specialized by-products, such as isolated molecules or the generation of nanoparticles with isolated compounds for use in various areas, these results may be limited by the lack of comparative studies that evaluate the total presence of compounds with biological activities in ahuilote compared to other unconventional fruits. However, considering the production system that does not require specialized inputs for fruit production and collection, the system becomes more resilient and capable of meeting sustainability criteria. Therefore, the application of a cluster model that includes local producers, companies, government, and academia working together to develop an appropriate model for the region (Fundación Biodiversidad, 2024) appears to be a suitable option for developing a bioeconomic model based on ahuilote.

6.4 Conclusions

This study identified that ahuilote (*Vitex mollis*), an unconventional fruit from the Valles region, Jalisco, Mexico, presents significant nutritional and bioactive characteristics, as well as a strong cultural attachment. It is described that its consumption has been restricted to Mexico since pre-Hispanic times, underscoring its importance in the identity and culinary tradition of the region. Additionally, it was observed that in Ahualulco de Mercado, Jalisco, there is an effort by the municipal government to revive the consumption of ahuilote through the Ahuilote Fair, where locals showcase their traditional ways of consumption while simultaneously generating regional development strategies. This suggests a local interest in promoting this unconventional fruit and conserving the cultural practices associated with it.

Given the results, ahuilote is part of the local circular bioeconomy on a smaller scale, but this does not mean it is of lesser importance. On the contrary, if we consider that other species are also exploited locally, this indicates an ecological resource utilization with traditional social implications and a natural development of the local bioeconomy. It is necessary to intensify and expand research to deepen the understanding of the nutritional and bioactive properties of ahuilote, to explore strategies for its promotion and utilization, as well as to create opportunities for research to evaluate the impact of fairs and other ahuilote promotion initiatives. This will help develop strategies that encourage its consumption and integration into the local circular bioeconomy.

Despite the ahuilote's bioeconomic potential, the inhabitants of the Valles de Jalisco region, particularly in Ahualulco de Mercado and Etzatlán, are not familiar with the concept of bioeconomy. This is due to the absence of a clear national policy on the subject in Mexico. Nevertheless, the ahuilote presents an opportunity to establish a local bioeconomic model that integrates traditional knowledge of its management, as well as its ecological and cultural significance. To achieve this, it is essential to foster innovation through education and training for local inhabitants in the development of new fruit-based products, while also encouraging collaboration with researchers to explore its potential medicinal benefits. In doing so, the ahuilote would not only strengthen the region's cultural identity but also drive a sustainable development model based on conservation and the responsible use of natural resources.

References

Acuña, R. (Ed.). (2017). Relaciones geográficas del siglo XVI: Nueva Galicia. Universidad Nacional Autónoma de México. http://ru.iia.unam.mx:8080/handle/10684/100

Alexandri, M., Christaki, S., Gkatzionis, K., Mourtzinos, I., & Tsouko, E. (2023). Residual biomass from major aromatic and medicinal flora of the Mediterranean: Challenges towards sustainable integration into food systems within the circular bioeconomy. *Trends in Food Science & Technology, 139*, 104123. https://doi.org/10.1016/j.tifs.2023.104123

Argueta, A., Sanabria Diago, O. L., Cano Contreras, E. J., & Medinaceli, A. (2018). Código de ética para la investigación etnobiológica en América Latina. *Ethnoscientia, 3*(2). https://doi.org/10.18542/ethnoscientia.v3i2.10239

Barros, F. B., De Sousa, F. F., De Andrade, J. P., Ramos, F. M., & Vieira-Da-Silva, C. (2021). Ethnoecology of miriti (*Mauritia flexuosa*, L.f.) fruit extraction in the Brazilian Amazon: Knowledge and practices of riverine peoples contribute to the biodiversity conservation. *Journal of Ethnobiology and Ethnomedicine, 17*(1). https://doi.org/10.1186/s13002-020-00430-z

Blancas, J., Abad-Fitz, I., Beltrán-Rodríguez, L., & Maldonado, B. (2023). El Coyotomate: Importante recurso aromático de las selvas secas. *La Jornada del Campo, 191*, 10. https://www.jornada.com.mx/2023/08/19/delcampo/articulos/coyomate-recurso-aromatico.html

Blanco, E. S., & Pérez Martínez, P. (2022). Conjunto especializado para la producción de instrumentos de obsidiana y el aprovechamiento de recursos lacustres dentro de un contexto isleño en el occidente mesoamericano. *Latin American Antiquity, 33*(3), 500–519. https://doi.org/10.1017/laq.2022.3

Bracco, S., Tani, A., Çalıcıoğlu, Ö., Gomez San Juan, M. & Bogdanski, A. (2019). *Indicators to monitor and evaluate the sustainability of bioeconomy. Overview and a proposed way forward.* Food and Agriculture Organization of the United Nations. Serial number 77. https://openknowledge.fao.org/items/620e2572-d3fb-478b-9fde-57bfec637731

D'Andrea, A., & D'Ulizia, A. (2023). Preserving local food traditions: A hybrid participatory approach for stimulating transgenerational dialogue. *Societies, 13*(4), 95. https://doi.org/10.3390/soc13040095

De la Cruz Borrego, J. G., & Caballero Rico, F. C. (2021). Bioeconomía: Una alternativa para México. Análisis bibliométrico a la Web of Science. *Revista Dilemas Contemporáneos: Educación, Política y Valores, 9*(1), Artículo 46.

Delgado-Vargas, F., Félix-Favela, F., Pio-León, J. F., López-Angulo, G., López-Valenzuela, J. A., Díaz-Camacho, S. P., & Uribe-Beltrán, M. d. J. (2010). Antibacterial activity and qualitative phytochemical analysis of *Vitex mollis* fruit. *International Journal of Green Pharmacy (IJGP), 4*(4) https://www.greenpharmacy.info/index.php/ijgp/article/view/162/183

Domínguez, A. (1997). La Laguna de Magdalena. *Estudios Jaliscienses, 29*, 16–30.

Estrada-Martínez, E., Vásquez-Dávila, M. A., Aldasoro, E. M., Sepúlveda-Varón, A. R., & Naranjo-Arcila, M. A. (2020). Metodologías participativas comunitarias en etnobiología. In F. Ruan Soto, A. Ramírez Terrazo, A. Montoya Esquivel, & R. Garibay Orijel (Eds.), *Métodos en etnomicología* (pp. 51–86). Instituto de Biología de la UNAM, GIDEM; Sociedad Mexicana de Micología.

Gomez San Juan, M., Harnett, S., & Albinelli, I. (2022). *Sustainable and circular bioeconomy in the climate agenda: Opportunities to transform agrifood systems*. FAO. https://doi.org/10.4060/cc2668en

Fundación Biodiversidad. (2024). The global bioeconomy. Preliminary Stocktake of G20 strategies and practices: A contribution to the Brazilian G20 presidency's global initiative on bioeconomy. Fundación Biodiversidad. https://ieeb.fundacion-biodiversidad.es/recursos/global-bioeconomy-preliminary-stocktake-g20-strategies-and-practices-contribution-brazilian

Gillani, S. W., Ahmad, M., Manzoor, M., Waheed, M., Iqbal, Z., Ullah, R., Pieroni, A., Zhang, L., Sulaiman, N., & Alrhmoun, M. (2024). The nexus between ecology of foraging and food security: Cross-cultural perceptions of wild food plants in Kashmir Himalaya. *Journal of Ethnobiology and Ethnomedicine, 20*(1). https://doi.org/10.1186/s13002-024-00721-9

Hernández Sampieri, R., Fernández Collado, C., & Baptista Lucio, P. (2010). *Metodología de la investigación*. McGraw-Hill Interamericana.

Instituto de Información Estadística y Geográfica de Jalisco (IIEG). (2022). *Ahualulco de Mercado: Diagnóstico del municipio*. Gobierno del Estado de Jalisco.

Instituto de Información Estadística y Geográfica de Jalisco (IIEG). (2023). Etzatlán: Diagnóstico del municipio. Gobierno del Estado de Jalisco.

Kaliora, A., & Dedoussis, G. (2007). Natural antioxidant compounds in risk factors for CVD. *Pharmacological Research, 56*(2), 99–109. https://doi.org/10.1016/j.phrs.2007.04.018

Martínez-Gordillo, M., Bedolla-García, B., Cornejo-Tenorio, G., Fragoso-Martínez, I., García-Peña, M. d. R., González-Gallegos, J. G., Lara-Cabrera, S. I., & Zamudio, S. (2017). Lamiaceae de México. *Botanical Sciences, 95*(4), 780–806. https://doi.org/10.17129/botsci.1871

Meza, L., & Rodríguez, A. (2022). Soluciones basadas en la naturaleza y la bioeconomía: Contribución a una transformación sostenible e inclusiva de la agricultura y a la recuperación pos-COVID-19. *Comisión Económica para América Latina y el Caribe (CEPAL).*. https://repositorio.cepal.org/bitstream/handle/11362/47886/1/S2200230_es.pdf

Montes-Avila, J., Ojeda-Ayala, M., López-Angulo, G., Díaz-Camacho, S., & Delgado-Vargas, F. (2018). Physicochemical properties and biological activities of melanins from the black-edible fruits *Vitex mollis* and *Randia echinocarpa*. *Food Measure, 12*, 1972–1980. https://doi.org/10.1007/s11694-018-9812-6

Montiel-Herrera, M., Camacho-Hernández, I. L., Ríos-Morgan, A., & Delgado-Vargas, F. (2004). Partial physicochemical and nutritional characterization of the fruit of *Vitex mollis* (Verbenaceae). *Journal of Food Composition and Analysis, 17*(2), 205–215. https://doi.org/10.1016/j.jfca.2003.09.001

Morales-Del Rio, J. A., Gutiérrez-Lomelí, M., Robles-García, M. A., Aguilar, J. A., Lugo-Cervantes, E., Guerrero-Medina, P. J., Ruiz-Cruz, S., Cinco-Moroyoqui, F. J., Wong-Corral, F. J., & Del-Toro-Sánchez, C. L. (2015). Anti-inflammatory activity and changes in antioxidant properties of leaf and stem extracts from *Vitex mollis* Kunth during in vitro digestion. *Evidence-Based Complementary and Alternative Medicine, 2015*, 1–8. https://doi.org/10.1155/2015/349235

Mountjoy, J. (2015). El Pantano y Otros Sitios del Formativo Medio en el Valle de Mascota, Jalisco. *Secretaría de Cultura, Gobierno de Jalisco*. https://doi.org/10.1017/S1045663500008099

National Research Council. (2008). *Lost crops of Africa* (Fruits) (Vol. III). The National Academies Press. https://doi.org/10.17226/11879

Orendain, T. E., & Michel, J. G. (2018). Paisaje biocultural y territorio, reflexiones sobre la gestión del patrimonio del "paisaje agavero y antiguas instalaciones del tequila". In *Libro de proceedings, CTV 2018. XII Congreso Internacional Ciudad y Territorio Virtual. "Ciudades y Territorios Inteligentes"* (pp. 5–7). CPSV.

Pennington, T. D., & Sarukhán, J. (2005). *Árboles tropicales de México. Manual para la identificación de las principales especies*. Universidad Nacional Autónoma de México/Fondo de Cultura Económica.

Perales, H. R., & Aguirre, J. R. (2008). Biodiversidad humanizada. In *Capital natural de México* (Conocimiento actual de la biodiversidad) (Vol. I). Conabio.

Pío-León, F., Montes-Ávila, J., Díaz-Camacho, S., & Delgado-Vargas, F. (2012). Biological activities and phytochemicals of the fruits of *Vitex* plants. In V. K. Gupta (Ed.), *Bioactive phytochemicals: Perspectives for modern medicine* (pp. 93–120). Daya Publishing House.

Rangel Cura, R. A., Zúniga González, C. A., Colón García, A. P., Losilla Solano, L., & Berrios-Zepeda, R. A. (2015). Medición de la contribución de la bioeconomía en América Latina: Caso Cuba. *Revista Iberoamericana De bioeconomía Y Cambio climàtico, 1*(1), 223–240. https://doi.org/10.5377/ribcc.v1i1.2150

Rodríguez, A. 2009. Metodología de la investigación etnozoológica. En: Medeiros Costa Neto E., D. Santos Fita, & M. Vargas Clavijo (Eds.), *Manual de Etnozoología* (253–272). Tundra Ediciones.

Salvador, R., Barros, M. V., Donner, M., Brito, P., Halog, A., & De, A. C. (2022). How to advance regional circular bioeconomy systems? Identifying barriers, challenges, drivers, and opportunities. *Sustainable Production and Consumption, 32*, 248–269. https://doi.org/10.1016/j.spc.2022.04.025

Sánchez Moreno, L. F. (2015). Recursos naturales en la región de los valles, Jalisco. *RIDE Revista Iberoamericana para la Investigación y el Desarrollo Educativo, 5*(10).

Taylor, S. J., & Bogdan, R. (2002). *Introducción a los métodos cualitativos de investigación*. Paidós.

Tetreault, D. V., & López, L. (2024). Diversidad biocultural en el estado de Jalisco: Pueblos indígenas y regiones de alto valor biológico. *Espiral (Guadalajara), 18*(51), 165–199. https://www.scielo.org.mx/scielo.php?script=sci_arttext&pid=S1665-05652011000200006

Vides-Rodríguez, C., González, R. S., Medina-Ortega, M. A., & Castañeda-Huizar, P. (2019). Las iniciativas de Desarrollo Local en el sector Agrícola. El caso de la región Valles, Jalisco. In Universidad Politécnica de San Luis Potosí (Ed.), *El papel de las empresas locales en el desarrollo regional de México* (pp. 249–278). Plaza y Valdés.

Uvario, K. M. L., Guardado, P. M., & Torreros, L. G. (2018). La economía naranja en el espacio rural: Análisis desde el desarrollo local en la Región de las Vías Verdes de los Valles, Jalisco. Redalyc.org. https://www.redalyc.org/articulo.oa?id=469554838002.

Zizumbo-Villarreal, D., & Colunga-García Marín, P. (2010). Origin of agriculture and plant domestication in West Mesoamerica. *Genetic Resources and Crop Evolution, 57*(6), 813–825. https://doi.org/10.1007/s10722-009-9521-4

Zuniga-González, C. A., López, M. R., Icabaceta, J. L., Vivas-Viachica, E. A., & Blanco-Orozco, N. V. (2022). Epistemología de la Bioeconomia. *Revista Iberoamericana De bioeconomía Y Cambio climático, 8*(15), 1786–1796. https://doi.org/10.5377/ribcc.v8i15.13986

Zuniga-González, C. A., Quiroga-Canaviri, J. L., Brambila-Paz, J. J., Ceballos-Pérez, S. G., & Rojas-Rojas, M. M. (2024). Frontier model of the environmental inefficiency effects on livestock bioeconomy [version 3; peer review: 3 approved, 1 not approved] F1000Research 2024, 11:1382. https://doi.org/10.12688/f1000research.128071.3

Zuniga-Gonzalez, C. A., Quiroga-Canaviri, J. L., Brambila-Paz, J. J., Ceballos-Pérez, S. G., & Rojas-Rojas, M. M. (2024). Formulation of an innovative model for the bioeconomy. *PLoS ONE, 19*(11), e0309358. https://doi.org/10.1371/journal.pone.0309358

Chapter 7
Analysis of the Bioeconomy Cost for *Sitotroga Cerealella: Trichogramma pretiosum* and *Chrysoperla externa*

Luis Francisco Moreno-Mayorga, Carlos Alberto Zuniga González, Silvio Javier Sáenz Rojas, and Walter Alberto Reyes Canales

Abstract Effective management of agricultural pests using biological controllers, such as *Sitotroga cerealella*, *Trichogramma*, and *Chrysoperla*, is crucial for sustainable pest control strategies. An in-depth cost analysis of these organisms is essential to optimize production processes and establish competitive pricing. Methodology: This study evaluated the cost structure associated with the production and reproduction of *Sitotroga cerealella*, *Trichogramma*, and *Chrysoperla*. Costs were categorized into raw materials, labor, and other overheads and converted into US dollars using an exchange rate of 36.7116. Results: The cost analysis revealed distinct cost patterns for each biological controller. For *Sitotroga cerealella*, the primary expense is the sterilization of *Sitotroga* eggs, accounting for 94% of the total costs (USD 1549.57), with sterilization of the wheat substrate and other stages contributing marginally. *Trichogramma* production represents the largest share of the total costs, making up 73% of the total (USD 7374.37), with breeder maintenance constituting the remaining 27%. In contrast, *Chrysoperla* production is heavily weighted toward larvae breeding, which accounts for 98% of the total costs (USD 517.41), with adult breeding contributing only 2%. Conclusions and Implications: The cost structure varies notably across the three biological controllers. To enhance cost-efficiency, focus should be directed toward optimizing the most expensive stages: improving sterilization processes for *Sitotroga*, refining production methods for *Trichogramma*, and exploring cost-saving measures in larvae breeding for

L. F. Moreno-Mayorga
Center for Research and Reproduction of Biological Controllers, National Autonomous University of Nicaragua, Leon, Nicaragua
e-mail: luis.moreno@ev.unanleon.edu.ni

C. A. Zuniga González (✉)
Research Center in Bioeconomy and Climate Change, National Autonomous University of Nicaragua, Leon, Nicaragua
e-mail: czuniga@ct.unanleon.edu.ni

S. J. Sáenz Rojas · W. A. Reyes Canales
National Autonomous University of Nicaragua, Leon, Nicaragua
e-mail: silvio.saenz@ev.unanleon.edu.ni; walter.canales@ev.unanleon.edu.ni

S. G. Ceballos Pérez et al. (eds.), *Sustainability in Latin America*,
https://doi.org/10.1007/978-3-032-09787-3_7

Chrysoperla. A comprehensive understanding of these cost dynamics facilitates better financial planning and supports the development of effective pricing strategies for biological pest control products.

Keywords Productivity analysis · Unit cost · Biological controllers · Sales · Organization

7.1 Introduction

In the western region of Nicaragua, extensive cotton cultivation caused significant alterations in the local ecosystem due to the excessive use of pesticides. This excess led to pests developing resistance to these products (Salamanca et al., 2023), resulting in the resurgence of secondary pests that previously had no economic relevance. Among these, the bollworm (*Helicoverpa zea*), the cotton leafworm *(Alabama argillacea*), and various species belonging to the genus *Spodoptera* (Falcon, 1971) stand out.

In response to the need for finding alternatives to the chemical control of these economically significant pests in Nicaragua, since 1982, the National Autonomous University of Nicaragua (UNAN-León) initiated research focused on *Trichogramma pretiosum* Riley (Hymenoptera: Trichogrammatidae), an egg parasitoid of lepidopterans, specifically of the family Noctuidae. Flanders (1959) pioneered the establishment of *Trichogramma* colonies using eggs of *S. cerealella* (Lepidoptera: Gelechiidae) as hosts. Subsequently, in other countries such as West Germany, Peru, and Colombia, techniques for the production of this parasitoid using the same host insect were developed. McLaren and Rye (1983) devised an efficient method for the mass rearing of *Trichogramma* wasps using eggs of *S. cerealella*. The ability to rear *T. chilonis* and other biological control agents on surrogate hosts such as *S. cerealella* has improved considerably (El-Hafez et al., 2001; Saljoqi et al., 2015).

In laboratory studies, Navarro and Zenner (1976) determined an average parasitism and emergence rate of *Trichogramma* of 83.7% and 78%, respectively, when analyzing the parasitism and emergence of this parasitoid under controlled conditions. De la Torre and Díaz (1982), also in laboratory settings, evaluated the control of the aromatic plant borer (*Corcyra cephalonica*, Lepidoptera: Galleriinae) using *Trichogramma oatmani*, achieving a 100% parasitism of the pest's eggs. Furthermore, De la Torre (1973) made significant discoveries regarding *Trichogramma oatmani* in his laboratory research.

He highlighted the exceptional capacity of this parasitoid for pest control, demonstrating its effectiveness by achieving 100% parasitism rates in eggs of various species of lepidopterans, specifically in situations of infestation by the aromatic plant borers (*Corcyra cephalonica*, Lepidoptera: Galleriinae). Additionally, he noted the parasitoid's ability to adapt and successfully reproduce using different host species, underscoring its potential as a biological control agent for agricultural pests.

Table 7.1 Pests controlled by *T. pretiosum* Riley, cultivation, and percentage of parasitism

Pests			
Common name	Scientific name	Crops	Parasitism (%)
Bellotero	*Helicoverpa zea*	Corn	85
Cucurbit fruit worm	*Diaphania* ssp.	Melon	65
Soybean worm	*Anticarsia gemmatalis*	Soy	60
Cane borer	*Diatraea saccharalis*	Sugarcane	100
False measuring locust	*Trichoplusia ni*	Bean	70
Cassava hornworm	*Manduca sexta*	Tomat0	100
Measuring lobster	*Alabama argillacea*	Cotton	95

Source: Cano Velázquez (2001)

These discoveries positioned *Trichogramma oatmani* as a promising tool in integrated pest management, demonstrating its high effectiveness and versatility in controlling harmful insects for crops. Bioeconomy has emerged as a crucial field in the quest for sustainable solutions for agricultural pest control. In particular, the Research Center for Production and Reproduction of Biological Controllers at the National Autonomous University of Nicaragua in León has stood out as a reference in this area. In this center, comprehensive work is carried out in the Entomophagous Laboratory, focused on the production and reproduction of *Sitotroga cerealella* as a parasitoid, with the aim of generating eggs of *Trichogramma* and *Chrysoperla* for sale. These biological control agents play a fundamental role in pest regulation, offering an effective and eco-friendly approach to reducing the incidence of diseases that affect crops.

The present study focused on a comprehensive cost analysis in the bioeconomy related to *Sitotroga cerealella*, specifically in the production of *Trichogramma* and *Chrysoperla*. This research is situated within the context of the growing importance of bioeconomy as an essential component for food security and agricultural sustainability. Through the detailed assessment of the costs involved in the production and commercialization of these biological control agents, the aim is to better understand their economic viability and potential impact on integrated pest management.

In Nicaragua, the mass production of the parasitoid was successful, allowing releases in multiple crops in León and Chinandega. These releases were conducted to evaluate the percentage of egg parasitism of various insect pests affecting economically important crops (see Table 7.1).

The results obtained demonstrated the effectiveness of the parasitoid, achieving up to 100% parasitism in eggs of *H. zea* in tomato crops during the period from 1990 to 2001 (see Table 7.1).

In the case of aphids on melon, the percentage of predation has been very significant, with control being 100% in most stations (Table 7.2) and the lowest being 60%.

This study is fundamental not only for its contribution to the efficient mass breeding of *Trichogramma* using *S. cerealella* eggs, but also for the significant expansion it offers to the field of commercialization of biological controllers. The

Table 7.2 Percentage of control by *C. externa* in aphids in melon crops

Aphids[a] Before release	After release	Stations	Predation control (%)
25	10	1	60
30	0	2	100
25	0	3	100
28	0	4	100
18	3	5	83
13	0	6	100

Source: Cano Velázquez (2001)
[a]23 aphids per square inch

ability to breed not only *T. chilonis*, but also *T. petrium* and *C. externa*, represents an important advance for the bioeconomy.

Improving mass breeding of these biological drivers opens the possibility for a more comprehensive cost analysis in the bioeconomy of *S. cerealella*, *T. petrium*, and *C. externa*. Production efficiency could have a significant impact on the economic viability of using these agents as pest control tools in agriculture. Evaluating the costs involved in their breeding, marketing, and their impact on the reduction of agricultural pests could provide essential perspective for decision-making at an economic and environmental level.

7.2 Literature Review

The economic evaluation of biological controllers in agriculture has been a topic of growing interest among researchers and experts in agricultural bioeconomics. Prominent authors in this field include Geoffrey K. Frampton, recognized for his work in the economic evaluation of biological controllers in agriculture, and R. Mark Norton, specialized in cost analysis in biological control (Frampton et al., 2000; Norton et al., 2016; Zúniga-González et al., 2022).

Frampton (1999, 2000) has provided valuable studies that examine in detail the costs associated with the production and application of biological controllers, highlighting the importance of considering not only the technical but also the economic aspects in the implementation of these strategies in agricultural systems. In a similar vein, Norton (1981) has addressed the economic analysis of biological controllers, exploring the costs of production, distribution, and the financial viability of their large-scale implementation. His work has highlighted the need to evaluate the economic effectiveness in relation to the biological effectiveness of these methods (Barrera et al., 2002; Mullen et al., 1997).

Other relevant experts such as Bruce E. Tabashnik and Roger Cook have contributed to the field through research that delves into the cost analysis of pest control strategies (Van der Putten et al., 2006). Tabashnik has explored the relationship between the costs of implementing biological controllers and pest resistance (Carrière et al., 2015; Tabashnik et al., 2003), while Cook has focused on evaluating the economic impacts of these methods in specific agricultural systems (Cooper, 1990; Cropper & Cook, 2000; Griffiths et al., 2000).

It also adds that the Center for Biological Control in Central America (CCBCA), created in 1989 and financed by the United States Agency for International Development (USAID), initially focused on classical biological control and the conservation of natural enemies. Over time, he shifted his focus to practical student teaching and commercial production of natural enemies for agricultural pest control. He stood out for discovering a new species, *Lixadmontia franki*, a parasitoid released for the biological control of the Mexican bromeliad weevil in Florida. In addition, it introduced and commercialized various biological controllers such as antagonistic fungi, predatory stink bugs, predatory mites, and entomophagous nematodes to control a variety of agricultural pests, while it has contributed to disseminating biological control knowledge through its graduates in Latin America (Cano, 1988; Cave et al., 2011).

Cueva (2012), in his research focused on promoting biological control in agriculture to support exports of fresh asparagus and avocados in La Libertad, Peru. He highlights its importance in pest management and its role in reducing costs, complying with international regulations and preserving the environment. The possibility for Peruvian agro-export companies to obtain sustainable competitive advantages and demonstrate social responsibility by adopting these practices is highlighted.

Finally, it is mentioned that SERVBIO, a company founded by the author between 1977 and 2008 in the Turbio Valley, Lara State, Venezuela, specialized in the production and application of biological controllers in Venezuelan agriculture. Initially, it focused on the control of sugarcane borers using the Amazonian fly *Lydella* (*Metagonistylum*) minense. Subsequently, starting in 1987, *Cotesia flavipes* was established and massively produced, managing to reduce the infestation intensity from 15% to 1.68% in the Azucarera Río Turbio area and reducing the cost of biological control from 41.25 to 1. The company ventured into the production of several biological controllers, including *Telenomus remus* for the control of *Spodoptera frugiperda* in corn and other grasses, as well as *Heterorhabditis bacteriophora* for the control of *Aenolamia varia* in sugarcane, achieving 80% control in Turen, Portuguesa state. This work has made it possible to have effective biological controllers for key pests in Venezuelan agriculture, while SERVBIO has contributed significantly to education and professional training through research, internships, theses, and consulting in the field of biological control (Ferrer Wurs, 2021).

7.3 Methodology

There is a growing need to apply cost accounting principles in entomophagous laboratory settings to efficiently evaluate the production and economic viability of rearing organisms such as *T. pretiosum* and *C. externa.* Adapting traditional cost principles to this specialized context is essential to understand the resources used and optimize the production of these biological controllers (Horngren et al., 2007; Ivoylov & Kochetov, 2021; Krześniak, 2022).

For cost analysis under laboratory conditions in the context of *T. pretiosum* and *C. externa*, cost accounting principles are applied to measure the resources consumed in the production and experimentation with these biological controllers (Krzesniak, 2023; Yardin, 2002).

Procedure Based on the stages defined in the reproduction process of *S. cerealella* as a host for the parasitization of *T. pretiosum* and *C. externa* (Park & Simpson, 2005; Walters, 1963). The basic cost formula is Eq. (7.1):

$$TC = FC + VC \tag{7.1}$$

where:

TC = Total cost
FC = Fixed cost
VC = Variables cost

Fixed costs are those that do not vary with production, such as laboratory rent, salaries of permanent staff, etc. Variable costs are directly related to production, such as specific inputs for the breeding and maintenance of *T. predariom* and *C. externa*, laboratory material, among others (Zuniga-Gonzalez, 2022; Zúniga-Gonzalez & Moreno-Mayorga, 2024).

The unit cost formula is useful for calculating the cost per unit produced (Eq. 7.2):

$$UC = TC \times UPC \tag{7.2}$$

where:

UC = Unit cost
TC = Total cost
UPC = Number of units produced

In the context of a biological controller production laboratory, each cost component must be considered in detail, from the personnel involved in breeding and maintenance, the inputs for the growth of the organisms, to the costs of research and development (Indzere et al., 2019).

Specific cost accounting methods, such as the standard costing method, can be applied to estimate and control costs in the production of *T. predariom* and *C. externa*. This method is based on establishing standards for the costs of the

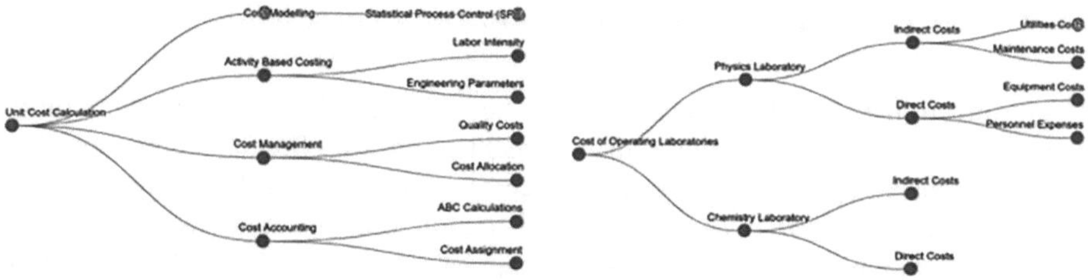

Fig. 7.1 Unit cost calculation and operating laboratories

resources used and then comparing these standards with the actual costs (Kubule et al., 2019). To evaluate production efficiency, cost indices such as the Average Unit Cost (Hadley, 1964) (Eq. 7.3) can be used:

$$AUC = \frac{TC}{TP} \tag{7.3}$$

where:

AUC = Average units cost
TC = Total cost
TP = Total production

These tools allow a detailed and rigorous analysis of the costs in the production of *T. pretiosum* and *C. externa* in a laboratory environment, facilitating decision-making and resource optimization in the production of these biological controllers (Taylor, 1923) (Fig. 7.1).

7.4 Results

In the field of biological control, accurate cost assessment is crucial for optimizing production processes and ensuring economic viability. This study provides a comprehensive analysis of the cost structure associated with the production of *Sitotroga cerealella*, a key biological control agent, and its byproducts *Trichogramma* and *Chrysoperla*. The analysis aims to identify and evaluate the costs incurred at different phases of the production process, offering insights into resource allocation and cost management strategies. By examining these costs, it can better understand the economic implications of scaling up production and commercializing these biological agents (El Defrawy & Said, 2022; Sultan et al., 2013; Zhang et al., 2020).

7.4.1 Cost Structure Analysis of *Sitotroga cerealella*

The production of *Sitotroga cerealella*, a key biological control agent, requires meticulous planning and resource management to ensure efficiency and economic viability. This study analyzes the cost structure associated with the production and reproduction of *Sitotroga cerealella*, focusing on the raw materials, labor, and operational expenses across various stages. Specifically, we examine the production process involving 5636 kg of wheat substrate, which is utilized to produce 18,788 grams of *Sitotroga cerealella* eggs, and 3758 grams of wheat infestation. By analyzing the costs across different phases, we aim to provide insights into resource allocation and cost management strategies, crucial for optimizing production and commercialization (Borzoui & Naseri, 2016).

Figure 7.2 suggests that, while these stages are necessary, they involve less resource intensity and thus lower costs compared to the egg sterilization stage. The minimal cost for adult extraction (0%) indicates it is the least expensive component of the production process. The disproportionate cost associated with the sterilization of *Sitotroga* eggs presents several implications for cost management (Kubule et al., 2019; Pitcher et al., 2002).

Given that egg sterilization comprises the majority of costs, efforts to optimize this process can have a significant impact on overall cost reduction. Consider exploring cost-saving technologies or methods that could enhance efficiency and reduce material or operational expenses associated with egg sterilization. Although other stages have relatively minor costs, maintaining efficiency across all stages is essential for comprehensive cost control. Regular reviews and optimizations can

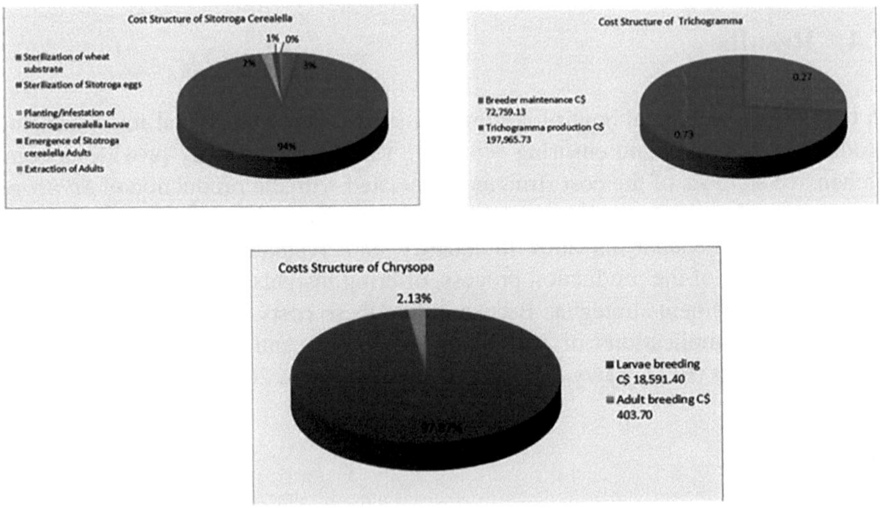

Fig. 7.2 Summary cost of *Sitotroga cerealella*, *Trichogramma*, and *Chrysoperla* production

prevent these smaller costs from accumulating or becoming significant (Ghaemmaghami et al., 2021).

7.4.2 Cost Structure Analysis of Trichogramma

Production *Trichogramma*, a key biological control agent, is produced through a multi-stage process involving significant investments in raw materials and labor. This study aims to analyze the cost structure associated with *Trichogramma* production, focusing on the various stages from breeder maintenance to final production. The objective is to provide a detailed breakdown of costs to identify high-expense areas and optimize resource allocation for enhanced economic efficiency (Cherif et al., 2021).

The production phase is significantly more costly than the breeder maintenance phase (Fig. 7.2). This high proportion highlights a potential area for cost optimization. Reviewing and refining production processes could lead to substantial cost savings. With the production phase being the most expensive, it is advisable to focus on improving efficiency and reducing costs in this area. This might involve optimizing resource usage, energy consumption, and operational processes. Understanding these costs is essential for evaluating the product's profitability, especially when comparing to potential selling prices in the market. Cost control in the production phase will directly impact profitability. Reducing costs in the production phase could significantly enhance profitability. Strategic measures to improve efficiency and lower expenses in this phase will be beneficial.

7.4.3 Cost Analysis for Chrysoperla Production

In this study, it analyzed the cost structure associated with the production of *Chrysoperla*, focusing on the breakdown of expenses into breeder maintenance and production phases. *Chrysoperla*, a beneficial insect used in biological pest control, involves various cost components throughout its production lifecycle. By examining the costs incurred at each stage, we aim to provide insights into the economic aspects of *Chrysoperla* production, which is essential for optimizing resource allocation and enhancing profitability.

The data indicates that the larvae breeding stage is the predominant cost component, accounting for 98% of the total expenditure. With a cost of USD 506.42, this stage is substantially more expensive compared to adult breeding. This significant cost allocation reflects the resource intensity and operational demands associated with breeding larvae, which likely involves extensive use of materials, space, and labor (Fig. 7.2).

In contrast, adult breeding constitutes only 2% of the total cost, amounting to USD 11.00. This relatively minor expenditure suggests that the resources and processes involved in adult breeding are less resource-intensive. This could be due to lower material costs or reduced labor requirements compared to larvae breeding.

Given the overwhelming cost dominance of larvae breeding, there are critical implications for cost management. To achieve cost reductions, a primary focus should be on optimizing larvae breeding processes. This may involve evaluating and improving the efficiency of resource use, reducing waste, and streamlining operational procedures. Implementing cost-saving measures in this stage can lead to substantial savings due to its significant cost share. Although the cost of adult breeding is relatively minor, it is still important to ensure that this stage operates efficiently. Monitoring and managing costs in this area can contribute to overall cost control and ensure that even minor expenses are kept in check.

7.5 Discussion

The analysis reveals that the most significant costs in the production of *Sitotroga cerealella* occur during the second stage, where the sterilization of eggs incurs substantial expenses. This is followed by costs in the third stage for seeding and infestation, reflecting the intensive resource requirements for maintaining optimal conditions. The costs for adult emergence and extraction are relatively lower, indicating that the production process is most resource-intensive during the sterilization and seeding phases (Kubule et al., 2019).

The data shows that sterilization of *Sitotroga* eggs constitutes the largest portion of the total cost, amounting to USD 1460.86 or 94% of the total expenditure. This substantial cost allocation highlights the critical importance and resource intensity of the egg sterilization process. The high cost could be attributed to the need for rigorous procedures to ensure the viability and quality of the eggs (Pitcher et al., 2002).

Given that egg sterilization comprises the majority of costs, efforts to optimize this process can have a significant impact on overall cost reduction. Consider exploring cost-saving technologies or methods that could enhance efficiency and reduce material or operational expenses associated with egg sterilization.

Understanding the cost distribution provides valuable insights into the financial management of *Sitotroga cerealella* production. By targeting cost reductions in the most significant expense areas, such as egg sterilization, the overall cost structure can be improved, potentially leading to better financial outcomes and pricing strategies (Ghaemmaghami et al., 2021).

The cost analysis reveals that the second stage, which involves *Trichogramma* production, is the primary driver of total costs. Raw materials and energy consumption are the major contributors to expenses in this stage. By contrast, the breeder maintenance stage represents a smaller proportion of the total cost but is essential for maintaining production continuity (Cherif et al., 2021).

The analysis reveals that *Chrysoperla* production incurs the most significant portion of the total cost, amounting to USD 66,699.31 or 76% of the overall expenditure. This substantial share indicates that the production phase is a major cost driver. The costs associated with raw materials, labor, and operational processes in this phase are key contributors to the high expenditure (Navarro & Zenner, 1976).

Breeder maintenance represents 24% of the total cost, which amounts to USD 21,437.43. While this is a smaller fraction compared to the production phase, it is still a critical component. Efficient breeder maintenance is necessary to ensure the availability of high-quality *Chrysoperla* for production. However, given its lower cost share, there may be limited scope for significant savings in this area compared to production (Koutsoula et al., 2023).

The high cost in the *Chrysoperla* production phase presents a notable opportunity for optimization. Strategies to improve efficiency, reduce waste, and lower production costs could lead to substantial financial benefits. For instance, reviewing and enhancing production processes, negotiating better rates for raw materials, and investing in more efficient technologies could contribute to cost reductions (Alghamdi et al., 2018).

Understanding the detailed cost structure is crucial for evaluating the economic viability of *Chrysoperla* production. By comparing production costs to market prices, businesses can assess their pricing strategies and profit margins. Effective cost management in the production phase can enhance profitability and ensure competitive pricing in the market.

To maximize profitability, it is advisable to focus on optimizing the production phase. Investments in process improvements, cost-effective resource management, and technological advancements should be prioritized. Additionally, regular cost monitoring and analysis will help in identifying further opportunities for cost reduction (Morgado et al., 2014).

7.6 Conclusion

The cost summary for *Sitotroga cerealella* production reveals that sterilization of eggs is the predominant cost driver, accounting for 94% of the total expenses. Focused efforts to optimize this stage can yield significant cost savings. Managing costs across all stages, even those with lower expenses, ensures a balanced approach to cost control and enhances overall financial performance.

The cost analysis reveals that *Trichogramma* production is the most expensive component, with a notable 73% of the total cost. Efforts to optimize this phase can lead to increased profitability and market competitiveness. Breeder maintenance, while less costly, remains crucial for the overall production process.

The cost summary for *Chrysoperla* production highlights that larvae breeding is the major cost driver, accounting for 98% of the total expenditure. Focusing on optimizing this stage is crucial for cost management and improving profitability. While adult breeding costs are minimal, maintaining efficiency in this stage also contributes to overall cost control. Effective cost management strategies will play a vital role in enhancing the financial performance of *Chrysoperla* production.

References

Alghamdi, A., Al-Otaibi, S., & Sayed, S. M. (2018). Field evaluation of indigenous predacious insect, Chrysoperla carnea (Steph.)(Neuroptera: Chrysopidae), fitness in controlling aphids and whiteflies in two vegetable crops. *Egyptian Journal of Biological Pest Control, 28*, 1–8. https://doi.org/10.1186/s41938-018-0026-3

Barrera, V. H., Quishpe, D., Crissman, C., Norton, G. W., & Wood, S. (2002). Evaluación económica de la aplicación de la tecnología de manejo integrado de plagas y enfermedades (MIPE) en el cultivo de papa en la Sierra del Ecuador.

Borzoui, E., & Naseri, B. (2016). Wheat cultivars affecting life history and digestive amylolytic activity of Sitotroga cerealella Olivier (Lepidoptera: Gelechiidae). *Bulletin of Entomological Research, 106*(4), 464–473. https://doi.org/10.1017/S000748531600016X. PMid:27019124.

Cano, E. (1988). *Cría Masiva y Liberación de Trichogramma pretiosum Riley con técnica mejorada en Nicaragua*. Tesis MSc. Nicaragua, Universidad Nacional Autónoma de Nicaragua.

Cano Velázquez, E. (2001). Manejo Integrado de Plagas. Avances en el Fomento de Productos Fitosanitarios No-Sintéticos. CATIE Costa Rica. Proyecto Producción y Aplicación masiva de Trichogramma y otros agentes biológicos. Universidad Nacional Autónoma de Nicaragua-León, Nicaragua. *Repositorio CATIE, reporte No, 60*, 93–96.

Carrière, Y., Crickmore, N., & Tabashnik, B. E. (2015). Optimizing pyramided transgenic Bt crops for sustainable pest management. *Nature Biotechnology, 33*(2), 161–168. https://doi.org/10.1038/nbt.3099. PMid:25599179.

Cave, R. D., Trabanino, R., & Pitty, A. (2011). Zamorano y sus contribuciones a la agricultura sostenible a través del control biológico de plagas. *Ceiba, 52*(1), 26–38. https://doi.org/10.5377/ceiba.v52i1.966

Cherif, A., Mansour, R., & Grissa-Lebdi, K. (2021). The egg parasitoids Trichogramma: from laboratory mass rearing to biological control of lepidopteran pests. *Biocontrol Science and Technology, 31*(7), 661–693. https://doi.org/10.1080/09583157.2020.1871469

Cooper, R. (1990). Cost classification in unit-based and activity-based manufacturing cost systems. *Journal of Cost Management, 4*(3), 4–14.

Cropper, P., & Cook, R. (2000). Activity-based costing in universities-five years on. *Public money and management, 20*(2), 61–68. https://doi.org/10.1111/1467-9302.00213

Cueva, F. D. (2012). El control biológico como estrategia para apoyar las exportaciones agrícolas no tradicionales en Perú: un análisis empírico. *Contabilidad y Negocios, 7*(14), 81–100. https://doi.org/10.18800/contabilidad.201202.006

De la Torre, S. M. (1973). *Estudios analíticos de la longevidad y fecundidad del Trichogramma fasciatum (Perkins) (Hymenoptera Trichogrammatidae) en función del alimento y número de huevos disponibles de Corcyra cephalonica (Stainton) (Lepidoptera, Galleriinae)*. Universidad de La Habana.

De la Torre, S. L., & Díaz, M. (1982). Posibilidad de controlar el bórer de las plantas aromáticas con el Trichogramma oatmani (Hymenoptera: Trichogrammatidae). Ciencia y Técnica en la Agricultura. Protección de Plantas.

El Defrawy, B., & Said, S. M. (2022). Predation efficiency and some biological aspects of the predator Chrysoperla carnea (Stephens) (Neuroptera: Chrysopidae) larvae feeding on different egg preys under laboratory conditions. *Menoufia Journal of Plant Protection, 7*(2), 19–27. https://doi.org/10.21608/mjapam.2022.222915

El-Hafez, A., et al. (2001). Acceptance and preference of pink bollworm and some lepidopterous eggs for parasitism by Trichogramma. *Egyptian Journal of Agricultural Research, 79*(2), 123–132.

Falcon, L. A. (1971). Microbial control as toll in integrated Control programs. In C. B. Huffaker (Ed.), *Biological Control* (pp. 346–362). https://doi.org/10.1007/978-1-4615-6528-4_15

Ferrer Wurs, F. (2021). Control biológico de plagas agrícolas en Venezuela: los logros históricos de la empresa Servicio Biológico (SERVBIO). *Revista de Ciencias Ambientales, 55*(1), 327–344. https://doi.org/10.15359/rca.55-1.16

Flanders, S. E. (1959). The employment of exotic entomophagous insects in pest control. *Journal of Economic Entomology, 52*(1), 71–75. https://doi.org/10.1093/jee/52.1.71

Frampton, G. K. (1999). Spatial variation in non-target effects of the insecticides chlorpyrifos, cypermethrin and pirimicarb on Collembola in winter wheat. *Pesticide Science, 55*, 885±886. https://doi.org/10.1002/(SICI)1096-9063(199909)55:9<875::AID-PS52>3.0.CO;2-S

Frampton, G. K. (2000). Large-scale monitoring of non-target pesticide effects on farmland arthropods in England. In J. J. Johnston (Ed.), *Pesticides and wildlife* (pp. 54–67). ACS Symposium Series. American Chemical Society. https://doi.org/10.1021/bk-2001-0771.ch005

Frampton, G. K., van den Brink, P. J., & Gould, P. J. L. (2000). Effects of spring precipitation on a temperate arable collembolan community analysed using principal response curves. *Applied Soil Ecology, 14*, 231–248. https://doi.org/10.1016/S0929-1393(00)00051-2

Ghaemmaghami, E., Fathipour, Y., Bagheri, A., Talebi, A. A., & Reddy, G. V. (2021). Quality control of the parasitoid wasp Trichogramma brassicae (Hymenoptera: Trichogrammatidae) over 45 generations of rearing on Sitotroga cerealella. *Insect Science, 28*(1), 180–190. https://doi.org/10.1111/1744-7917.12757. PMid:31990456.

Griffiths, B. S., Ritz, K., Bardgett, R. D., Cook, R., Christensen, S., Ekelund, F., et al. (2000). Ecosystem response of pasture soil communities to fumigation-induced microbial diversity reductions: an examination of the biodiversity-ecosystem function relationship. *Oikos, 90*(2), 279–294. https://doi.org/10.1034/j.1600-0706.2000.900208.x

Hadley, G. (1964). A comparison of order quantities computed using the average annual cost and the discounted cost. *Management Science, 10*(3), 472–476. https://doi.org/10.1287/mnsc.10.3.472

Horngren, C. T., Foster, G., & Datar, S. M. (2007). *Contabilidad de costos un enfoque gerencial. Décimo segunda edicíón*. Pearson educacíón.

Indzere, Z., Melvere, M., Muizniece, I., & Blumberga, D. (2019). The evaluation of factors affecting bioeconomy development using transdisciplinary approach. *Environmental and Climate Technologies, 23*(3), 360–369. https://doi.org/10.2478/rtuect-2019-0101

Ivoylov, O. O., & Kochetov, A. G. (2021). A method for analyzing the typical structure of direct costs in the cost of a laboratory test. *Klinicheskaia Laboratornaia Diagnostika, 66*(3), 187–192. https://doi.org/10.51620/0869-2084-2021-66-3-187-192. PMid:33793120.

Koutsoula, G., Stamkopoulou, A., Pekas, A., Wäckers, F., Broufas, G., & Pappas, M. L. (2023). Predation efficiency of the green lacewings Chrysoperla agilis and C. mutata against aphids and mealybugs in sweet pepper. *Bulletin of Entomological Research, 113*(2), 162–168. https://doi.org/10.1017/S0007485322000426. PMid:36082696.

Krutyakova, V., Molchanova, O., Bulgakov, V., & Rucins, A. (2020). Experimental investigation of ways to increase efficiency of breeding grain moth. *Engineering for Rural Development, 19*, 1168–1174. https://doi.org/10.22616/ERDev.2020.19.TF283

Krześniak, G. (2022). Production cost calculation approach based on data collected during manufacturing process. *System Safety: Human-Technical Facility-Environment, 4*(1), 19–27. https://doi.org/10.2478/czoto-2022-0003

Krzesniak, G. (2023). Effective data usage for the proper and beneficial automotive production cost improvement. *Journal Of Achievements In Materials and Manufacturing Engineering, 119*(1), 27–34. https://doi.org/10.5604/01.3001.0053.8696

Kubule, A., Indzere, Z., & Muizniece, I. (2019). Modelling of the bioeconomy system using interpretive structural modelling. *Agronomy Research, 17*(4), 1665–1678. https://doi.org/10.15159/AR.19.170

McLaren, I. W., & Rye, W. J. (1983). The rearing, storage and release of Trichogramma ivelae Pang and Chen (Hymenoptera: Trichogrammatidae) for control of Heliothis punctiger Wallengren (Lepidoptera: Noctuidae) on tomatoes. *Australian Journal of Entomology, 22*(2), 119–124. https://doi.org/10.1111/j.1440-6055.1983.tb01855.x

Morgado, L. N., Resendes, R., Moura, M., & Ventura, M. A. M. (2014). Pollen resources used by Chrysoperla agilis (Neuroptera: Chrysopidae) in the Azores, Portugal. *European Journal of Entomology, 111*(1), 143–146. https://doi.org/10.14411/eje.2014.015

Mullen, J. D., Norton, G. W., & Reaves, D. W. (1997). Economic analysis of environmental benefits of integrated pest management. *Journal of Agricultural and Applied Economics., 29*(2), 243–253. https://doi.org/10.1017/S1074070800007756

Navarro, A., & Zenner, I. (1976). Estudios básicos tendientes a mejorar el uso de Trichogramma sp. en el control integrado de plagas en Colombia. *Revista Colombiana De Entomología, 2*(1), 13–24. https://doi.org/10.25100/socolen.v2i1.10404

Nordlund, D. A., & Morrison, R. K. (2021). Mass rearing of Chrysoperla species. In *Advances in insect rearing for research and pest management* (pp. 427–439). https://doi.org/10.4324/9780429043246-28

Norton, G. W. (1981). The productivity and allocation of research: US agricultural experiment stations, revisited. *North Central Journal of Agricultural Economics*, 1–12. https://doi.org/10.2307/1349403

Norton, M. R., Malinowski, D. P., & Volaire, F. (2016). Plant drought survival under climate change and strategies to improve perennial grasses. A review. *Agronomy for Sustainable Development, 36*, 1–15. https://doi.org/10.1007/s13593-016-0362-1

Park, J., & Simpson, T. W. (2005). Development of a production cost estimation framework to support product family design. *International Journal of Production Research, 43*(4), 731–772. https://doi.org/10.1080/00207540512331311903

Pitcher, S. A., Hoffmann, M. P., Gardner, J., Wright, M. G., & Kuhar, T. P. (2002). Cold storage of Trichogramma ostriniae reared on Sitotroga cerealella eggs. *BioControl, 47*, 525–535. https://doi.org/10.1023/A:1016533116798

Salamanca, L., Echeverri-Rubiano, C., Rondón, M., Manzano, M. R., Sarmiento, C. E., Michaud, J. P., & Vargas, G. (2023). Divergent life histories between wild and commercially-sourced Cotesia flavipes (Hymenoptera: Braconidae) parasitizing four species of Diatraea (Lepidoptera: Crambidae). *Biological Control, 178*, 105144. https://doi.org/10.1016/j.biocontrol.2022.105144

Saljoqi, A. R., Muhammad, G., Huma, Z., Ahmad, B., Zada, H., Rehman, S., et al. (2015). Screening of various irrigated wheat varieties against Angoumois grain moth, Sitotroga cerealella (Olivier)(Lepidoptera: Gelechiidae). *Journal of Entomology and Zoology Studies, 3*, 36–41.

Sultan, R., Khan, J., Haq, E., Mehmood, T., Khattak, R. Z., & Akhtar, N. (2013). Biological parameters of Trichogramma chilonis Ishii (Trichogrammatidae: Hymenoptera) feeding on Sitotroga cerealella eggs at three constant temperatures. *Pakistan Journal of Agricultural Research, 26*(1).

Tabashnik, B. E., Carrière, Y., Dennehy, T. J., Morin, S., Sisterson, M. S., Roush, R. T., et al. (2003). Insect resistance to transgenic Bt crops: lessons from the laboratory and field. *Journal of Economic Entomology, 96*(4), 1031–1038. https://doi.org/10.1093/jee/96.4.1031. PMid:14503572.

Taylor, J. S. (1923). A statistical theory of depreciation: Based on unit cost. *Journal of the American Statistical Association, 18*(144), 1010–1023. https://doi.org/10.1080/01621459.1923.10502133

Van der Putten, W. H., Cook, R., Costa, S., Davies, K. G., Fargette, M., Freitas, H., et al. (2006). Nematode interactions in nature: models for sustainable control of nematode pests of crop plants? In *Advances in Agronomy* (Vol. 89, pp. 227–260). https://doi.org/10.1016/S0065-2113(05)89005-4

Walters, A. A. (1963). Production and cost functions: An econometric survey. *Econometrica: Journal of the Econometric Society*, 1–66. https://doi.org/10.2307/1910949

Yardin, A. (2002). Una revisión a la teoría general del costo. *Revista Contabilidade & Finanças, 13*, 71–80. https://doi.org/10.1590/S1519-70772002000300006

Zhang, N., Xie, L., Wu, X., Liu, K., Liu, C., & Yan, Y. (2020). Development, survival and reproduction of a potential biological control agent, Lasioseius japonicus Ehara (Acari: *Blattisociidae*), on eggs of Drosophila melanogaster (Diptera: *Drosophilidae*) and Sitotroga cerealella (Lepidoptera: *Gelechiidae*). *Systematic and Applied Acarology, 25*(8), 1461–1471. https://doi.org/10.11158/saa.25.8.8

Zuniga-Gonzalez, C. A. (2022). Stages of the bioeconomy of the production process *Sitotroga cerealella*. protocols.io. https://doi.org/10.17504/protocols.io.bp2l69x1klqe/v1

Zúniga-Gonzalez, C. A., & Moreno-Mayorga, L. F. (2024). Data for analysis of the bioeconomy cost of Sitotroga cereallela: Trichogramma pretiosum and Chrysopa externa. *Mendeley Data, V2*. https://doi.org/10.17632/f8wtw4hmmh.2

Zúniga-González, C. A., López, M. R., Icabaceta, J. L., Vivas-Viachica, E. A., & Blanco-Orozco, N. V. (2022). Epistemology of bioeconomy. *Iberoamerican Journal of Bioeconomy and Climate Change, 8*(15), 1786–1796. https://doi.org/10.5377/ribcc.v8i15.13986

Chapter 8
Composting Training Program: A Collaborative Response that Contributes to the Circular Bioeconomy of Organic Solid Waste

Juan Camilo Fontalvo-Buelvas, Yadeneyro de la Cruz-Elizondo, Anabell Rosas-Domínguez, and Karla Beatriz Jerezano-Díaz

Abstract Composting is a biological process of great utility in the sustainable use of solid waste. This biotechnological innovation contributes to the circular bioeconomy of waste in urban areas. However, there are instances where citizens lack the technical capabilities to conduct these initiatives efficiently. In this sense, the objective of this study was to highlight the collaborative initiative between the City of Xalapa and Universidad Veracruzana (UV) in which citizens are offered the Virtual Composting Training Program. This qualitative study employed a descriptive and interpretative approach in which the narrative biographical method was used to define the object of study and collect, analyze, and report the data. Among the results, it is worth highlighting that the training program was held six times, with the participation of 219 people. A total of 58.5% of the participants satisfactorily completed the training process and provided evidence of the development of homemade composts in which they had begun to transform organic solid waste. This governance experience demonstrates the value of collaborative alliances between academia, government agencies, and communities in advancing toward a circular bioeconomy. Undoubtedly, citizen education for the bioeconomy can facilitate the separation of waste at source and help to develop capacities for its sustainable use.

J. C. Fontalvo-Buelvas (✉)
Universidad Nacional Autónoma de México, Escuela Nacional de Estudios Superiores, Unidad Morelia, Morelia, Michoacán, Mexico
e-mail: jfontalvo@iies.unam.mx

Y. de la Cruz-Elizondo
Universidad Veracruzana, Facultad de Biología, Xalapa, Veracruz, Mexico
e-mail: ydelacruz@uv.mx

A. Rosas-Domínguez
El Colegio de Veracruz, Xalapa, Veracruz, Mexico

K. B. Jerezano-Díaz
Plataforma Metropolitana de Formación en Agroecología, Xalapa, Veracruz, Mexico

S. G. Ceballos Pérez et al. (eds.), *Sustainability in Latin America*,
https://doi.org/10.1007/978-3-032-09787-3_8

Keywords Education for the bioeconomy · Governance · Organic waste · Sustainability · Linkage

8.1 Introduction

Bioeconomy is an emerging concept that is becoming increasingly important in academic research and political arenas. This paradigm is conceived as a technical solution to the dominant economic model and a primary means of achieving sustainable development (Aguilar et al., 2019). The bioeconomy includes the production of biological materials with renewable characteristics, as well as their transformation into value-added products, such as bio-based inputs or bioenergy (Barañano et al., 2021). Currently, bioeconomy is approached from an interdisciplinary perspective and presents three dominant areas of focus: biotechnology, biomaterials science, and bioecology (Bugge et al., 2019). In the case of the latter, there is a clear need to integrate traditional knowledge related to ecological processes to optimize the use of energy and nutrients, promote biodiversity, and improve soil health.

Composting is one of the most prominent eco-techniques for transforming organic solid waste into fertilizer. The process is based on the natural decomposition and recycling of organic matter in terrestrial ecosystems (Epstein, 2019). In recent years, the large-scale implementation of composting has begun in order to take advantage of the significant volume of organic waste generated in cities. In Mexico, the most important cases of this type are developed in Mexico City, where there are seven composting plants in current operation, with another five planned for the future (Secretaría de Medio Ambiente, 2021). However, one of the most emblematic facilities in the southeast of the country is the Composting Center of Xalapa, the capital of Veracruz state. This socio-ecological innovation was the result of a collaborative effort between local government, academia, the private sector, and civil society during the 2018–2021 administration (de la Cruz-Elizondo et al., 2019). The facility processes 30 tons of organic waste daily from markets, flower shops, restaurants, and the wholesale market.

The facility was part of a comprehensive program of the Directorate of Environment and Sustainability of the City of Xalapa. This program was not limited to the collection and final disposal of urban solid waste generated in the municipality. Indeed, a key element of the strategy was the implementation of effective waste management practices to minimize the input of waste into landfills and reduce greenhouse gas emissions (Gautam & Agrawal, 2021). For this, additional orchards and composting modules were established in community management centers, educational institutions, and housing areas. The objective was to reduce the entry of waste into the municipal landfill, generate an adequate and efficient management program, incorporate citizens in the separation and use processes, and maintain the capacities of the Composting Center (Ayuntamiento de Xalapa, 2020a). Recognizing the shared responsibility between the government and citizens in the integrated management of urban waste (Seow et al., 2022), the Department of Agroecology and

Environmental Education was established in 2018. This administrative department was created to promote a culture of environmental care and conservation, and the adoption of sustainable practices in daily life through initiatives such as composting, sustainable food production, and rainwater harvesting (Ayuntamiento de Xalapa, 2020b).

Given the favorable collaborative environment that emerged during this administration, the City of Xalapa worked together with the University of Veracruz to deliver the Virtual Garden Training Program (Rosas-Domínguez et al., 2023). However, participants often expressed the need to expand the sessions related particularly to the practice of composting and vermicomposting, leading to the development of a specialized course in response. Likewise, users highlighted the difficulties of accessing fertile soil in the city since paving covers the majority of the area. In this context, the objective of this study is to highlight the collaborative approach adopted by the University of Veracruz and the City of Xalapa in offering the Virtual Composting Training Program to the citizens. This initiative provided theoretical and practical tools to implement homemade composts in the context of the coronavirus (COVID-19) pandemic, supporting food production and strengthening urban agriculture in the city of Xalapa in Veracruz, Mexico.

8.2 Methods

This is a qualitative research study of a descriptive and interpretative type, in which the characteristics of a given event are detailed and the essential significance of the phenomenon is analyzed and inferred (Elliott & Timulak, 2021). The study employed the narrative biographical method, a technique usually used to analyze the personal, professional, and social aspects through the lens of testimonies provided by the subjects (Eichsteller, 2019). In this instance, the method proved effective for documenting the most relevant aspects of the experience, taking into account the discourse of the training session facilitators and participants. The methodological strategy comprised four distinct stages: selection of the topic or object of study, data collection, analysis, and reporting (Fig. 8.1).

The focus of this research is an analysis of the collaboration between Universidad Veracruzana and the City of Xalapa to offer citizens the Virtual Composting Training Program. Information was gathered from the autobiographical accounts of three facilitators and one of the organizers, as well as a review of official documents related to the didactic planning of the program and its sessions. Furthermore, surveys were applied to participants in different editions of the program to obtain information on age, sex, origin, and reasons for enrollment. Participant observation was employed during the virtual sessions. This technique consists of the systematic and non-intrusive collection of data in environments of social interaction between the researcher and the observed subjects (Lareau, 2021) and comprised both the training period (December 2020–November 2021) and 1 year of follow-up (to November 2022).

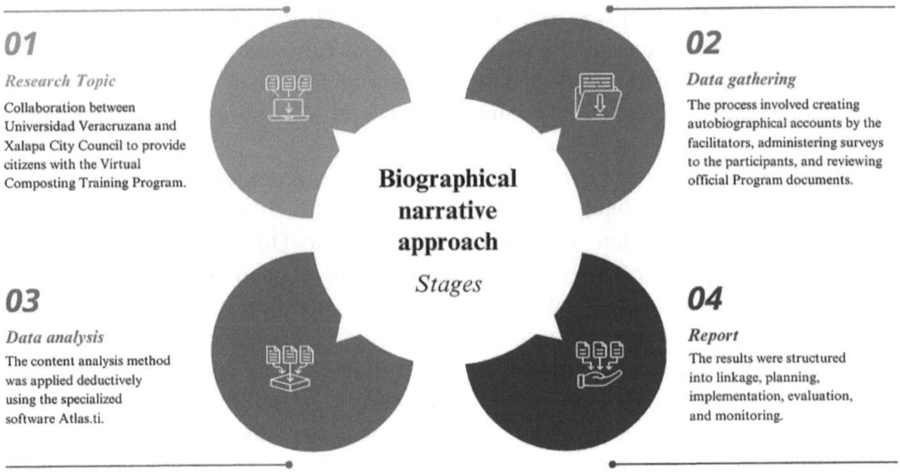

Fig. 8.1 Four stages of the narrative biographical method employed to document the experience. (Source: own production)

The data analysis stage was conducted using the content analysis technique, a qualitative research tool used to draw inferences from the systematic and objective identification of specific characteristics found within the discourses (Lindgren et al., 2020). This process was conducted deductively using the Atlas.ti software (v. 24.0.1), taking as a reference the following thematic categories: linkage between the parties, collective planning of the training program, implementation, learning and program assessment, and tracking of experiences. Finally, the information was reported in chronological order following the aforementioned thematic categories.

8.3 Results

8.3.1 Linkage Between Government and Academy

The Virtual Composting Training Program was developed from an extension of the non-formal link between the City of Xalapa and Universidad Veracruzana, which had already been implementing the Virtual Training Program in Orchards. Specifically, the collaboration exists between the Agroecological Garden of the Faculty of Biology—Xalapa Region and the Department of Agroecology and Environmental Education, which is part of the Directorate of Environment and Sustainability. The coordinator of the Agroecological Garden participated, along with three graduate students who were developing their master's theses featuring topics that were closely related to the program. The theses in question are: "The Agroecological Garden and its sustainable processes to support the substantive functions of the Faculty of Biology-Xalapa, Veracruz," "Integral management of organic solid waste: alliance between the Central de Abastos, farmers, and the Composting Center of Xalapa (Veracruz)," and "Vermicomposting as an alternative for the management of orange peel waste in Martínez de la Torre, Veracruz." Two technicians from the Department of Agroecology and Environmental Education (DAEA) and some invited members of the Subdirectorate of Integral Waste Management and the Xalapa Composting Center participated, representing the City Council staff. This collaborative effort began in 2020 with an informal partnership. However, a collaboration agreement was established in 2022 between the relevant parties to continue with these and other activities of mutual interest.

8.3.2 Collective Planning

The instructional design of this program was developed collectively between the parties, using the ADDIE model as a reference point. This tool includes five phases: analysis, design, development, implementation, and evaluation. It is an interactive instructional design process, characterized by its simplicity and flexibility in application. The organizing group, comprising representative members from the Agroecological Garden and the Department of Agroecology and Environmental Education, adopted the tool to facilitate reflective dialogues, taking into account the needs initially raised by users of the Garden Program. Figure 8.2 provides a summary of how the ADDIE paradigm was applied to the planning of the Virtual Composting Training Program.

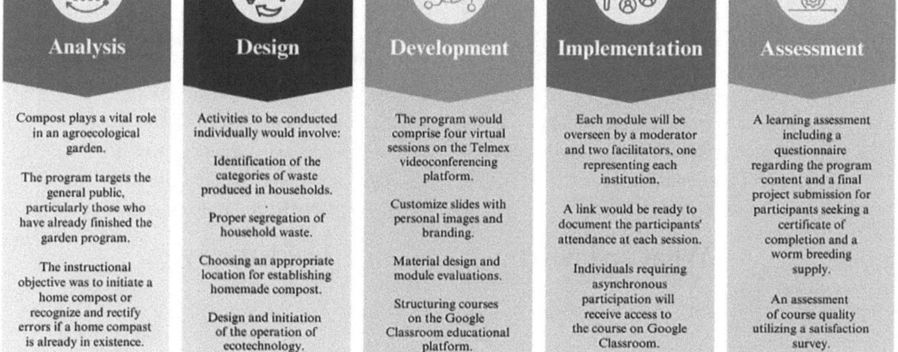

Fig. 8.2 Initial application of the ADDIE model for the instructional design of the Virtual Composting Training Program. (Source: own production)

8.3.3 Program Implementation

The program had six virtual editions from December 2020 to November 2021. The work methodology included an oral presentation for each module, via a PowerPoint presentation. All sessions were recorded and uploaded to the Google Classroom platform. Access was granted to the participants to view the recordings of the sessions and the facilitators' presentations. Complementary learning material was also provided, including infographics, guides, videos, and manuals to facilitate in-depth exploration of the topics. A detailed overview of the contents of each module is presented in Fig. 8.3.

A total of 798 people applied for enrollment in the program. Of these, 73.8% were between 20 and 39 years of age. Applications were received from nine minors between 13 and 17 years of age and the oldest person who registered was 70 years old. In terms of gender, 71.0% were women, 28.5% were men, and 0.5% identified as "other." It should be noted that only 27.7% of the participants resided in Xalapa, while the majority of those interested came from other places, such as Coatepec and

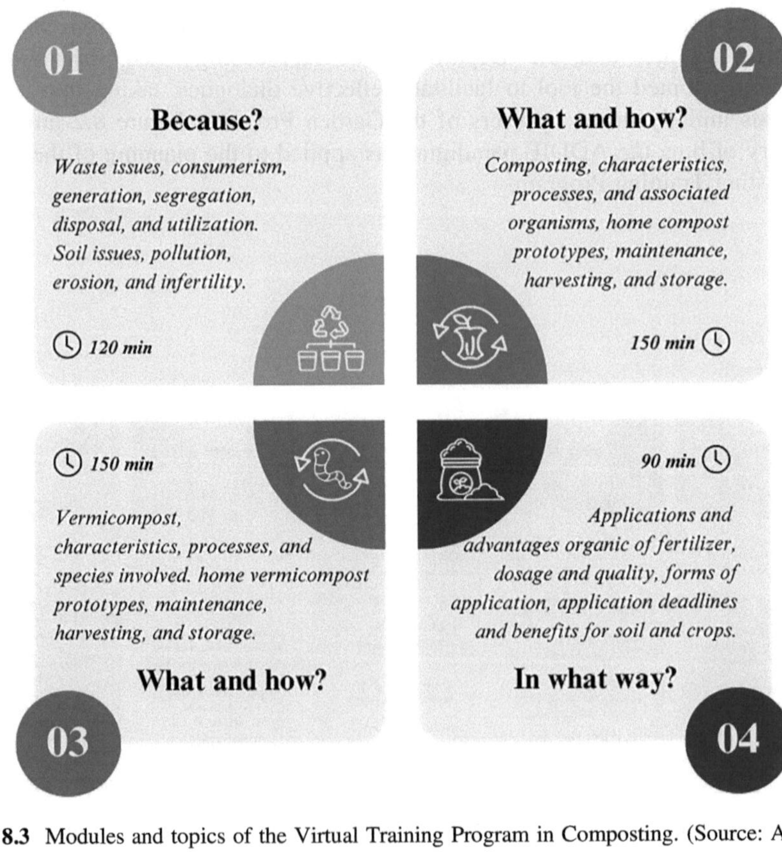

Fig. 8.3 Modules and topics of the Virtual Training Program in Composting. (Source: Author's own production)

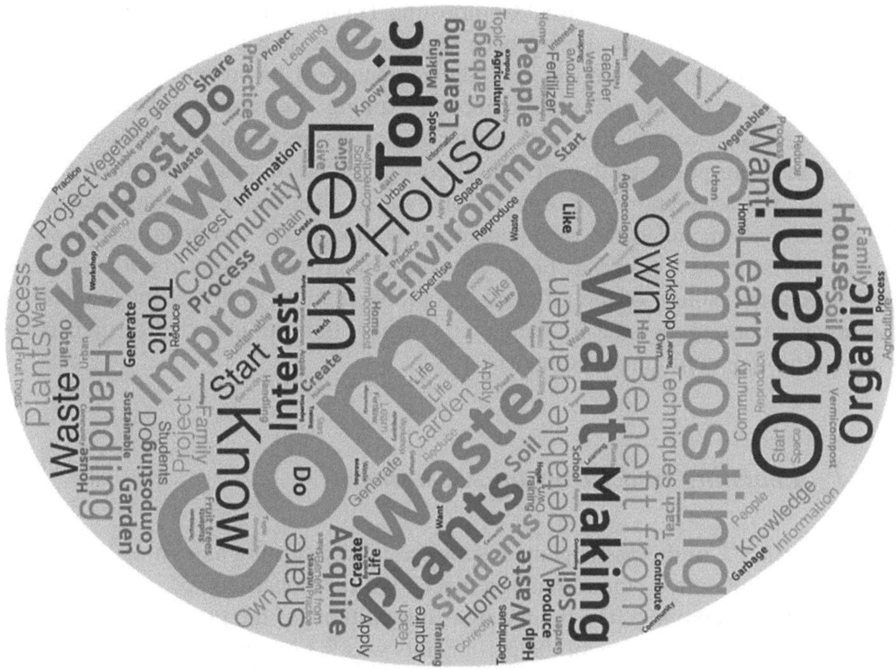

Fig. 8.4 Frequent word cloud (8 to 275 mentions) of the reasons for applying for registration to the Virtual Composting Training Program. (Source: Author's own production)

Perote, Veracruz state, and the states of Chiapas, Oaxaca, Tabasco, Estado de México, Puebla, and Mexico City, as well as various places in Peru, Colombia, El Salvador, Costa Rica, Uruguay, and Paraguay. Finally, the reasons given for enrolling in the course were to learn or improve learning about composting or vermicomposting processes, to better manage solid waste generated at home, and to complement horticulture or gardening projects. A word cloud generated from the 798 reasons recorded for applying for enrollment in the Program is presented in Fig. 8.4.

A total of 219 people were accepted and participated in the editions of the training program. Initially, up to 50 people were accepted per edition but, thanks to the increase in the capacity of the virtual platform, up to 100 participants were accepted for each of the last two editions. However, the actual attendance at the sessions was much lower, with an average participation of 36.5 people per edition. Of the 219 participants, 73.0% were female, 26.5% were male, and one person (0.5%) identified their gender as "other." In terms of participant age, the data are concentrated in the 20 to 29 and 30 to 39 age ranges, representing 71.4% of the total. The lowest frequency was observed in the extreme ranges, with 2.8% of participants under 20 years of age and 1.9% of 60 years or older.

Unlike the preventive approach of the Virtual Garden Training Program, where the importance of critical consumption to reduce the amount of waste was emphasized at all times, the Virtual Composting Training Program was based on the assumption that waste has already been generated and that proper management is necessary to minimize its impact and take advantage of its processing through the production of compost and vermicompost. Among the topics addressed in the training sessions, composting and vermicomposting received the most attention because they are practices that fit into almost any type of space and allow for the recycling of a large amount of organic solid waste. The most common concerns of the participants included the maintenance of compost and vermicompost in small spaces and the most appropriate prototypes for both processes. It is important to note that, during the edition of the program, people were motivated to manage not only the organic waste generated by their own consumption but also that of their community, through community composts.

8.3.4 Learning and Program Assessment

A total of 58.5% of the participants responded to the learning assessment questionnaires and submitted their final project, which consisted of improving or implementing a compost. During the first four editions of the program, the average score attained in evaluations of the participants was 7.3 out of 10 possible points. In these first editions of the program, it was observed that participation was decreasing, so it was decided to apply the questionnaires separately at the end of each module. Thus, the average result of the evaluations of the last two programs was 8.6 points. The questions that were most often answered incorrectly were those related to basic concepts about waste, the ability of worms to feed, and the identification of green manures.

On the other hand, 60.6% of the participants responded to the quality evaluation of the Program. Most of the respondents felt that the syllabus was complete and well structured, with sufficient and up-to-date content. More than 80% of the respondents reported learning new topics. The people who led the sessions were all rated favorably, with technical and didactic skills ranging from high to very high. The reported perception was very good for dissemination of the program, punctuality, and the virtual platforms used. The issue that elicited the lowest ratings (41.6%) was the limited number of sessions in which to address the topics, and some participants requested more practical experiences. The overall evaluation of the program was very good and 98% of the participants stated that they would recommend the courses offered by the Agroecological Garden of the Faculty of Biology and the Department of Agroecology and Environmental Education. Among the most common comments were suggestions for additional topics to be addressed in future workshops and the suggestion to increase the time or number of sessions in which to adequately cover the contents.

8.3.5 Tracking Experiences

Monitoring of progress and resolution of doubts of the participants was achieved through direct attention to the requests received through personal WhatsApp messages. This platform allowed for quick and efficient communication, facilitating one-on-one interaction and ongoing support. In addition, the WhatsApp groups that had been previously created in the Virtual Garden Training Program were used, reinforcing the already integrated learning communities. However, in order to improve follow-up in similar projects implemented in the future, it is recommended to consider new additional strategies to maintain interest and sharing of experiences, for example, creating WhatsApp groups exclusive to each topic, assigning facilitators or guides to each group, organizing periodic meetings to present experiences, and hosting guided tours to learn about successful projects.

Finally, to motivate the participants to initiate their own vermicomposting projects, those who lived in Xalapa were offered the opportunity to receive a worm inoculum (500 g of worm compost containing approximately 100 worms) from the Xalapa Composting Center. The DAEA, in coordination with the Composting Center, donated a total of 194 worm breeding feet to participants in this and other activities aimed at promoting the reduction and proper management of solid waste generated in the local municipality.

8.4 Discussion

This citizen training experience was made possible by the collaborative link between academia and the local government to address the incipient social awareness of the problem of organic solid waste generation. In this sense, there are few experiences of government-academia linkage to develop training programs on composting. Most of the documented initiatives have been developed by local governments, such as the training program of the Home Composting Project that seeks to strengthen environmental education in the Oreamuno canton in Costa Rica (Solano-Araya, 2023), although some training processes have emerged as university extension projects, such as the workshops on composting and vermicomposting developed in schools by the National University of La Plata (Palancar & Lanfranco, 2013). One of the most interesting strategies, however, is the “Ba composta” program proposed for the city of Buenos Aires (Argentina). This is a very complete program that includes the installation of homemade compost, training, materials for raising awareness, accompaniment, economic incentives, funding, and monitoring and evaluation using indicators (Gugliotta, 2022).

One remarkable aspect of this experience is the great interest and response of citizens to learn and share knowledge about eco-technologies such as composting. In this regard, it is important to note that 73% of the interest in this case was presented by women, who have a broad sense of care, both human and environmental. Indeed, a recent study in Indonesia showed that the active role of women has a notable impact on the correct separation of household waste and its efficient use through composting (Hamidiana et al., 2021). More specifically, in the field of bioeconomy and other alternative economies such as ecological economics, ecofeminism problematizes and critiques hegemonic economies (Kongar & Berik, 2021). More recently, it has been reported that a social vision involving feminist frameworks, strategies, and tools is emerging as a path to sustainability, highlighting the role of women as potential key actors in the transition to the bioeconomy (Sanz-Hernández et al., 2022). Although these novel epistemological frameworks are currently restricted to the academic field, their opening to civil society is a matter of time and civic education can be an optimal tool for this objective.

In this sense, we believe that education is a tool for social transformation that can contribute to the formation of eco-citizens with the appropriate skills to reduce irresponsible consumerism and implement eco-techniques for waste utilization (Cho & Park, 2023). Therefore, just as environmental education has evolved toward education for sustainability, we believe that an "education for the bioeconomy" should be promoted in parallel. Such an education strategy would allow society to learn about this paradigm, its dimensions, advantages, and importance. In particular, it must be practical and offer specific tools for people to see how they can get involved and take a leading role. This is essential to enable the transition to learning communities and educational societies with full awareness to manage solidarity ecology and its shared responsibility in current socio-ecological problems (Mathevet et al., 2018). In this regard, the home composting plan with house-to-house awareness campaigns implemented in the municipality of Neapoli-Sykies (Greece) is prominent. This was a successful education strategy that demonstrated broad awareness among citizens, who were motivated by environmental advice and concern (Karkanias et al., 2016).

It should be noted that citizen education processes provide detailed instruction to the public about institutions, government agencies, and their principles and values, including the actions they take in terms of community well-being (Zepke, 2013). In this case, the Virtual Composting Training Program not only provided technical capabilities but also publicized the importance of the Xalapa Composting Center. Together with the training strategy, this far-reaching government activity can inspire society to generate new circular bioeconomy initiatives at multiple scales and with different benefits. For example, home composting, which can save on the purchase of fertilizer for the garden; community composting, which encourages neighborhood cohesion and collective action; school composting, which promotes social awareness, as well as sustainable habits and lifestyles; and commercial composting as a new venture to produce organic fertilizers for the agricultural sector. In this way, the bioeconomy of organic waste can contribute to the incorporation of nutrients into the soil, increasing its fertility and making farmers less and less dependent on chemically synthesized fertilizers (Cidón, 2024). Recently, a study in Vietnam showed that training programs in municipal composting influence the decision of citizens to participate in home composting schemes (Takahashi et al., 2019).

We therefore propose education for the bioeconomy as a key tool that can provide information, co-produce knowledge, and generate learning experiences around the bioeconomy as a science, social movement, daily practice, and agroecology (Astier et al., 2017). In this sense, people who are informed and educated about the bioeconomy could be equipped with the skills to exercise and defend their environmental rights and responsibilities in society and play an active role. Indeed, citizen education stimulates social participation, an essential aspect of governance processes and the generation of public policies (Chanda, 2024). Although the bioeconomy has an important impact on the political sphere, it is essential to develop training strategies that involve civil society in particular (Rodríguez & Prestvik, 2020). Citizens can make a difference if they are aware of "the power of the consumer" in the economic sphere, but also of "the power of social organization" in the democratic and political sphere. Shared governance should therefore stimulate greater acceptance of the bioeconomy to harness its potential in the context of sustainable development (Böcher et al., 2020). However, it is necessary to take into account the fact that this path, like those of the other alternative economies that struggle to confront neoliberal economies, will present multiple challenges.

8.5 Conclusions

The Virtual Composting Training Program was a successful experience of collaboration between the University of Veracruz and the City of Xalapa, which provided theoretical-practical tools to promote home composting in the context of the COVID-19 pandemic. Under this scenario of crisis, people confined to their homes became more sensitive about the relationship between economic growth, ecosystem health, and human well-being. This also happened at a historical moment when universities closed their doors, but through virtual platforms became more aware of the social responsibility they have towards society. A similar situation occurred with governments, which began to socialize their good environmental practices and engage citizens in similar actions from home.

The main contribution of this study is the proposed set of steps that can be replicated by other institutions to generate links and promote spaces for citizen education. This experience shows that local governments and academia can work together to address socio-ecological problems. Likewise, intersectoral links can provide citizens with tools for social participation in the circular bioeconomy of organic solid waste. Undoubtedly, alliances and education for the bioeconomy are paramount to creating ideal ecosystems of socio-ecological innovation and achieving Sustainable Development Goals. However, it is advisable to continue to strengthen links through legal structures that allow the continuity of collaboration, regardless of administrative changes. Similarly, it is essential to integrate training programs into more far-reaching public policies to achieve comprehensive waste management at the regional level. This could contribute to the development of sustainable communities and cities, capable of reducing their emissions, and with unique characteristics of resilience to the impact of climate change.

References

Aguilar, A., Twardowski, T., & Wohlgemuth, R. (2019). Bioeconomy for sustainable development. *Biotechnology Journal, 14*(8), 1800638. https://doi.org/10.1002/biot.201800638

Astier, M., Argueta, J. Q., Orozco-Ramírez, Q., González, M. V., Morales, J., Gerritsen, P. R., & González-Esquivel, C. (2017). Back to the roots: Understanding current agroecological movement, science, and practice in Mexico. *Agroecology and Sustainable Food Systems, 41*(3–4), 329–348. https://doi.org/10.1080/21683565.2017.1287809

Ayuntamiento de Xalapa. (2020a). Apropiación social de los procesos selectivos de los residuos sólidos de carácter orgánico de mercados y centros comerciales de la ciudad de Xalapa, Veracruz México. *RIGL: Revista Iberoamericana De Gobierno Local,* (16), 8. https://dialnet.unirioja.es/servlet/articulo?codigo=9428455

Ayuntamiento de Xalapa. (2020b). *Manual Específico de Organización. Dirección de Medio Ambiente y Sustentabilidad.* Gaceta Municipal, número extraordinario 045. https://cloud.xalapa.gob.mx/index.php/s/XdNiEygJ9xbLRwp

Barañano, L., Garbisu, N., Alkorta, I., Araujo, A., & Garbisu, C. (2021). Contextualization of the bioeconomy concept through its links with related concepts and the challenges facing humanity. *Sustainability, 13*(14), 7746. https://doi.org/10.3390/su13147746

Böcher, M., Töller, A. E., Perbandt, D., Beer, K., & Vogelpohl, T. (2020). Research trends: Bioeconomy politics and governance. *Forest Policy and Economics, 118*, 102219. https://doi.org/10.1016/j.forpol.2020.102219

Bugge, M. M., Hansen, T., & Klitkou, A. (2019). What is the bioeconomy? In A. Klitkou, A. M. Fevolden, & M. Capasso (Eds.), *From waste to value: Valorisation pathways for organic waste streams in circular bioeconomies* (pp. 19–50). Taylor & Francis. https://doi.org/10.4324/9780429460289

Chanda, C. T. (2024). Civic education and citizen participation in local governance: A case of Lusaka District, Zambia. *International Journal of Research Publication and Reviews, 5*(3), 4628–4637. https://ijrpr.com/uploads/V5ISSUE3/IJRPR23961.pdf

Cho, S. M., & Park, M. J. (2023). Analysis of eco-citizenship contents elements in home economics textbooks for the introduction of ecological transformation education. *Journal of Korean Home Economics Education Association, 35*(2), 1–20. https://doi.org/10.19031/jkheea.2023.6.35.2.1

Cidón, C. F. (2024). Bioeconomy for sustainable organic farming. In V. K. Garg & N. Kataria (Eds.), *Bioeconomy for sustainability* (pp. 189–204). Springer. https://doi.org/10.1007/978-981-97-1837-5_8

de la Cruz-Elizondo, Y., Martínez-Tlapa, F., Fontalvo-Buelvas, J. C., & Narave-Flores, H. V. (2019). Gestión social de la Planta de Lombricompostaje en el municipio de Teocelo, Veracruz. In C. J. D. Covarrubias, J. L. Capdevila, & Z. M. A. Chamorro (Eds.), *De la gestión social al desarrollo local. Debates en torno a procesos de gestión social* (pp. 237–262). Universidad de Guadalajara. http://www.publicaciones.cucsh.udg.mx/kiosko/2019/accion_comunitaria.pdf

Eichsteller, M. (2019). There is more than one way–a study of mixed analytical methods in biographical narrative research. *Contemporary Social Science, 14*(3–4), 447–462. https://doi.org/10.1080/21582041.2017.1417626

Elliott, R., & Timulak, L. (2021). *Essentials of descriptive-interpretive qualitative research: a generic approach* (Essentials of Qualitative Research). American Psychological Association. https://doi.org/10.1037/0000224-000.

Epstein, E. (2019). The science of composting. CRC Press. https://doi.org/10.1201/9780203736005.

Gautam, M., & Agrawal, M. (2021). Greenhouse gas emissions from municipal solid waste management: A review of global scenario. In S. S. Muthu (Ed.), *Carbon footprint case studies. Environmental footprints and eco-design of products and processes* (pp. 123–160). Springer. https://doi.org/10.1007/978-981-15-9577-6_5

Gugliotta, A. (2022). *Programa "Ba composta": promoción al compostaje domiciliario para una ciudad sustentable* [Master's thesis]. Universidad Nacional de San Martín. http://ri.unsam.edu.ar/handle/123456789/2565

Hamidiana, Z., Heddy, S., Wafi, I. A., Sari, A. N., & Fauziah, H. (2021). The effectiveness of women's empowerment in aerobic composting. *Environmental Engineering & Management Journal (EEMJ), 20*(7), 1047–1056. http://www.eemj.icpm.tuiasi.ro/pdfs/vol20/no7/Full/1_154_Hamidiana_20.pdf

Karkanias, C., Perkoulidis, G., & Moussiopoulos, N. (2016). Sustainable management of household biodegradable waste: Lessons from home composting programmes. *Waste Biomass Valor, 7*, 659–665. https://doi.org/10.1007/s12649-016-9517-1

Kongar, E., & Berik, G. (Eds.). (2021). The Routledge handbook of feminist economics. Routledge. https://doi.org/10.4324/9780429020612.

Lareau, A. (2021). *Listening to people: A practical guide to interviewing, participant observation, data analysis, and writing it all up.* University of Chicago Press. https://eprints.lse.ac.uk/114007/

Lindgren, B. M., Lundman, B., & Graneheim, U. H. (2020). Abstraction and interpretation during the qualitative content analysis process. *International Journal of Nursing Studies, 108*, 103632. https://doi.org/10.1016/j.ijnurstu.2020.103632

Mathevet, R., Bousquet, F., Larrère, C., & Larrère, R. (2018). Environmental stewardship and ecological solidarity: Rethinking social-ecological interdependency and responsibility. *Journal of Agricultural and Environmental Ethics, 31*, 605–623. https://doi.org/10.1007/s10806-018-9749-0

Palancar, T., & Lanfranco, J. (2013). Reducción de residuos orgánicos urbanos. Capacitación pública sobre compostaje y lombricompostaje. *Extensión en red,* (4), 1–5. https://perio.unlp.edu.ar/ojs/index.php/extensionenred/article/view/1685/1662

Rodríguez, D. G. P., & Prestvik, A. S. (2020). The need for stakeholder engagement and participative governance to promote bioeconomy. In U. S. Nagothu (Ed.), *The bioeconomy approach* (pp. 211–236). Routledge. https://doi.org/10.4324/9780429320651

Rosas-Domínguez, A., Jerezano-Díaz, K. B., Fontalvo-Buelvas, J. C., & de la Cruz-Elizondo, Y. (2023). Programa virtual de capacitación en huertos: una vinculación entre el ayuntamiento de Xalapa y la Universidad Veracruzana en México. In M.Á. Piñar Álvarez, X. del Alba León Estrada & A. Wojtarowski Leal (coords.), *Reflexiones y experiencias en torno al desarrollo regional sustentable* (pp. 361–377). El Colegio de Tlaxcala & El Colegio de Veracruz. https://revistacoltlax.mx/omp/index.php/repositoriocoltlax/catalog/book/65

Sanz-Hernández, A., Jiménez-Caballero, P., & Zarauz, I. (2022). Gender and women in scientific literature on bioeconomy: A systematic review. *Forest Policy and Economics, 141*, 102762. https://doi.org/10.1016/j.forpol.2022.102762

Secretaría de Medio Ambiente. (2021). *Programa de gestión integral de residuos de la Ciudad de México 2021-2025.* Gobierno de la Ciudad de México. https://bit.ly/3LpSA6a.

Seow, T. W., Guan, L. W., Mohamed, S., Radzuan, I. S. M., & Abas, M. A. (2022). Good governance practices in recycling Programmes. *Environment and Ecology Research, 10*(2), 260–266. https://doi.org/10.13189/eer.2022.100215

Solano-Araya, C. N. (2023). *Propuesta de diseño de un programa de capacitación para el Departamento de Higiene y Ornato Ambiental de la Municipalidad de Oreamuno a través de un análisis del proyecto ambiental de compostaje doméstico en el 2023* [Bachelor's thesis]. Instituto Tecnológico de Costa Rica. https://hdl.handle.net/2238/14479

Takahashi, Y., Nomura, H., & Yabe, M. (2019). Modeling home composting behavior toward sustainable municipal organic waste management at the source in developing countries. *Resources, Conservation and Recycling, 140*, 65–71. https://doi.org/10.1016/j.resconrec.2018.08.016

Zepke, N. (2013). Lifelong education for subjective well-being: How do engagement and active citizenship contribute? *International Journal of Lifelong Education, 32*(5), 639–651. https://doi.org/10.1080/02601370.2012.753125

Chapter 9
Sustainable Innovation: Nanocellulose Extraction with Natural Deep Eutectic Solvents in Lignocellulosic Residues in Latin American

Susana Rojas-Varela, Juliana Morales-Castro, Ana M. Ávila-Galván, Orlando A. Manzanares-Meza, and María I. Guerra-Rosas

Abstract Currently, Latin America generates a significant volume of waste classified as lignocellulosic, as these materials are composed of lignin, cellulose, and hemicellulose. From such waste, valuable and innovative products can be derived, including nanocellulose (CNC), a cellulose-based material at the nanoscale with diverse industrial applications. Acid hydrolysis, the conventional method for CNC production, although is highly efficient, presents various disadvantages such as the use of strong acids, the generation of toxic waste, and high energy demands. As a sustainable alternative, green solvents, as those Natural Deep Eutectic Solvents (NADES), have emerged as a promising approach. These solvents are biodegradable, non-toxic, and highly effective in dissolving lignocellulosic components, enabling environmentally friendly and competitive processes for nanocellulose production. Among the lignocellulosic residues, some agro-industrial by-products have been studied for nanocellulose obtention and presented in this chapter, such as banana peels, sugarcane bagasse, rice husk waste, pineapple leaves, garlic skin, agave bagasse, and leaves where conventional methods were used. Therefore, a big potential exists to explore the use of green solvents like NADES in nanocellulose recovery in conjunction with other green technologies including ultrasound and microwave technologies in greening nanocellulose extraction in Latin America where the use of green solvents in lignocellulosic residues for valorization still represents a big opportunity for knowledge generation.

Keywords Nanocellulose · Lignocellulosic Residues · NADES · Green solvents

S. Rojas-Varela · J. Morales-Castro (✉) · A. M. Ávila-Galván · O. A. Manzanares-Meza · M. I. Guerra-Rosas
Tecnológico Nacional de México/Instituto Tecnológico de Durango, Depto. de Ing. Quim. y Bioq. LANAEPBI, Durango, Dgo, México
e-mail: 23040245@itdurango.edu.mx; jmorales@itdurango.edu.mx; 13040786@itdurango.edu.mx; 13040814@itdurango.edu.mx; m.guerra@itdurango.edu.mx

S. G. Ceballos Pérez et al. (eds.), *Sustainability in Latin America*,
https://doi.org/10.1007/978-3-032-09787-3_9

9.1 Introduction

Globally, the population growth has caused a direct outgrowth on food demand, which, as result, has impacted the amount of food residues generated. These residues can be differentiated in different types according to the stage of the food production process where they are generated. Lignocellulosic residues come from the stages of harvesting, and industrial processing, and are generally composed by agricultural waste (straws, stubbles, bagasses, peels, and stems), forestry residues (wood branches, bark, leaves, sawdust, and shavings), and some biomasses produced by industrial activities (residual pulp and bagasses) that usually do not receive a treatment and represent a potential danger to the environment (Arevalo-Gallegos et al., 2017; Devi et al., 2022).

In 2022, global production of lignocellulosic residues reached 2.3 billion tons, highlighting the urgent need for valorization alternatives. This trend is evident in developing countries, particularly in Latin America. Brazil leads with an annual production of 430 million tons, primarily from sugarcane bagasse and other agricultural residues. Argentina follows with 100 million tons, Colombia with 72 million tons, Mexico with 60 million tons, and Chile with 4.6 million tons of lignocellulosic residues, which include corn residues, forestry biomass, and more (FAO, 2023; Matiacevich et al., 2023; Peñaranda-Gonzâlez et al., 2017). These figures underscore the significance of developing sustainable strategies for utilizing these abundant resources.

Lignocellulosic residues are mainly composed of cellulose, hemicellulose, and lignin, structural components which present an excellent potential for chemical and biofuels production in a sustainable way (Devi et al., 2022). The large quantities of lignocellulosic residues produced and the chemical composition of it represent an opportunity area to develop processes and value-added products.

9.2 Cellulose

Lignocellulosic biomass (LCB) is considered a renewable resource due to its natural growth and constant regeneration, making it readily available (Ojo, 2023). It has a complex structure mainly composed of cellulose, hemicellulose, and lignin, along with minor compounds like proteins, lipids, ashes, and minerals (Roy et al., 2020). Cellulose is the primary component, typically comprising 15–40% of LCB, followed by hemicellulose (20–35%) and lignin (15–40%) (Roy et al., 2020; Scapini et al., 2021). These components are interconnected through intra- and interpolymeric linkages such as ether, ester, and hydrogen bonds (Roy et al., 2020).

Cellulose, the most abundant natural polymer on Earth, is the main structural component of natural fibers and is valued for its high biocompatibility, biodegradability, renewability, and hydrophilicity, making it suitable for a wide range of applications (Chen et al., 2017). It is a high-molecular-weight linear biopolymer with a semi-crystalline structure, comprising both amorphous and crystalline fractions. Its structure is composed of D-glucose units linked by β-1,4 glycosidic bonds, forming cellobiose units, which are two linked D-glucose units whose union lead to form cellulose chains (Heinze, 2016; Roy et al., 2020).

The cellobiose units contain three hydroxyl groups that strongly interact with adjacent glucose units, creating intra- and intermolecular hydrogen bonding networks. This bonding leads to tightly packed crystalline regions in cellulose fibrils, resulting in fibers with low solubility in water and high resistance to organic solvents and chemicals within the plant cell wall. Conversely, the amorphous regions in the cellulose structure, which have a lower degree of order, contribute to the flexibility of LCB (George & Sabapathi, 2015). These crystalline characteristics give cellulose fibers exceptional mechanical and physical properties (Phanthong et al., 2018).

The amorphous regions in cellulose, being more accessible to chemical and enzymatic treatments, facilitate the isolation of elementary fibrils known as nanocellulose, which are widely recognized as polymeric reinforcements with numerous applications (Jiang et al., 2021).

9.3 Nanocellulose

Cellulose molecules with a size between 1 and 100 nm are considered as nanocellulose, a biodegradable nanofiber with great potential for applications. Nanocellulose exhibits notable chemical, structural, mechanical, biological, and physicochemical properties that makes it a tunable material, such as highlightable mechanical strength, biocompatibility, optical transparency, biodegradability, high specific surface area, and polyfunctionality (Huang et al., 2019).

9.3.1 Chemical Structure

The fractionation of cellulosic fibers into nanocellulose produces two main types: cellulose nanocrystals (CNC) and cellulose nanofibers (CNF). CNC are rigid, nanometric particles (<100 nm) with high crystallinity due to their highly ordered structure, offering high specific strength and surface area. Their morphology varies depending on the cellulose source (George & Sabapathi, 2015). CNF, similar in diameter (10–100 nm), can be much longer, reaching up to millimeters. They have both amorphous and crystalline regions, providing greater flexibility compared to CNC (Nechyporchuk et al., 2016). CNC and CNF are typically produced through different extraction methods, such as cooking, bleaching, and acid hydrolysis, resulting in variations in the type of nanocellulose obtained (Phanthong et al., 2018).

9.3.2 *Extraction Methods*

Nanocellulose extraction methods present as main objective the size reduction of cellulosic fibers, through the implementation of different mechanical processes, such as high-pressure homogenization, microfluidization, liquefaction; enzymatic processes (mainly by enzymatic hydrolysis); and chemical processing, such as TEMPO oxidation and acid hydrolysis (Fig. 9.1), being acid hydrolysis the most common method applied for nanocellulose production (George & Sabapathi, 2015).

9.3.2.1 Acid Hydrolysis

Acid hydrolysis is a widely used method for producing nanocellulose due to its high efficiency in generating CNC. The process leverages the structural differences in cellulose fibers, targeting the amorphous regions with strong acids (H_2SO_4 or HCl) to break β-1,4 glycosidic bonds. This selective hydrolysis leaves behind the crystalline regions with a high crystallinity index (Phanthong et al., 2018). The process involves several steps (Fig. 9.2): (a) reacting cellulose with the acid under controlled conditions (temperature, time, acid concentration, and solid-liquid ratio); (b) halting the hydrolysis with water; (c) sonication to form stable nanocellulose suspensions; and (d) drying the suspension to obtain solid nanocellulose (Wulandari et al., 2016).

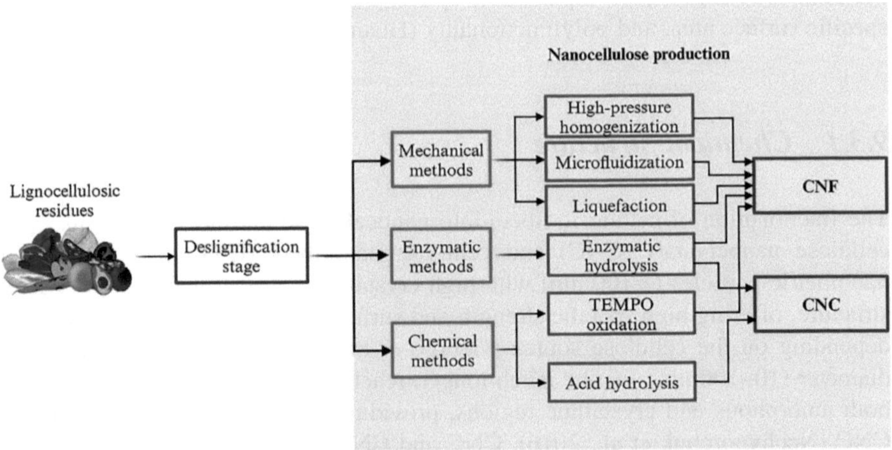

Fig. 9.1 Nanocellulose production methods

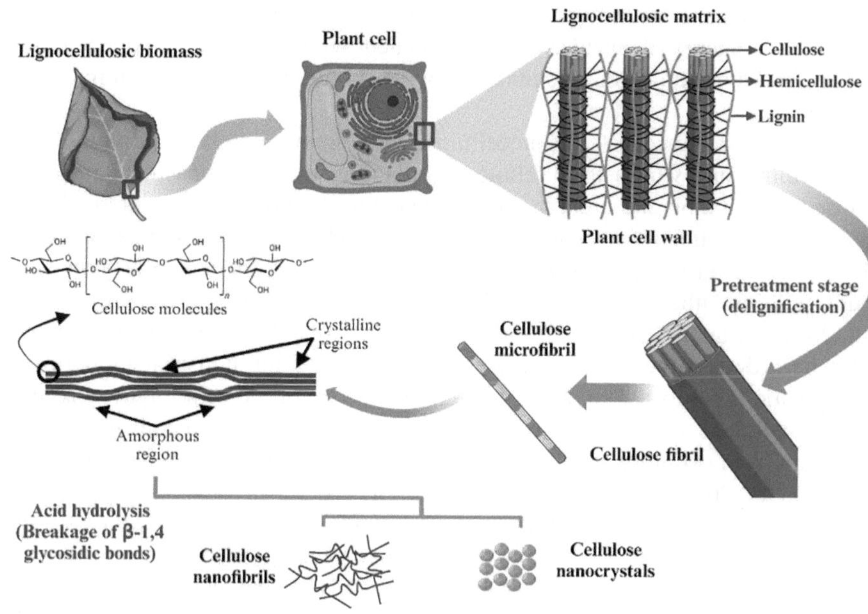

Fig. 9.2 Nanocellulose production by acid hydrolysis from lignocellulosic residues process

9.4 Green Technologies for Nanocellulose Extraction

Although acid hydrolysis is the typical nanocellulose production method, it generates toxic and hazardous effluents that are often inadequately disposed of, particularly in developing countries. This practice poses severe environmental risks by polluting soil and water ecosystems, making it inconsistent with the principles of Green Chemistry. Green Chemistry, a concept introduced in the 1990s, advocates for designing products and processes that minimize or eliminate hazardous substance use and generation (Anastas & Warner, 1998; Horvath & Anastas, 2007). The 12 principles of Green Chemistry, now globally adopted, aim to establish "green processes" by promoting cleaner synthetic routes, renewable resources, alternative reaction conditions, and the design of less toxic and biodegradable chemicals (Murcia Fandiño & Esquiaqui Marín, 2021).

Implementing these principles not only supports more sustainable production processes but also enhances the quality of life for workers, fosters innovation in environmentally friendly technologies and promotes a circular economy. These efforts yield economic and social benefits, encouraging a responsible chemical industry committed to environmental preservation (Pino, 2020).

In this context, alternative methods for nanocellulose production have been explored since the 1950s, including ultrasound treatments (Morehead, 1950), mechanical treatments (Pereira & Arantes, 2018), enzymatic hydrolysis (Bondancia et al., 2017), and the use of alternative solvents like Deep Eutectic Solvents (DES) and Natural DES (NADES) (Almeida et al., 2023). These green technologies typically do not generate hazardous waste or toxic byproducts. However, challenges such as high energy demands, costs associated with DES and NADES liquor recovery, and the acquisition of suitable enzymes hinder their widespread industrial adoption.

Ultrasound treatments for nanocellulose production involve applying high-intensity ultrasonic waves to a cellulose-containing medium, causing bubbles to collapse by cavitation and breaking down the cellulosic fibers (Santos et al., 2014). Other mechanical treatments, such as high-shear and high-pressure homogenization, ball milling, fine grinding, and steam explosion, similarly generate shear forces that disrupt cellulose fibers on a nanoscale (Almeida et al., 2023; Hernández-Varela et al., 2021; Furtado et al., 2021; Cherian et al., 2010). While these methods do not require hazardous solvents, their high energy demands limit industrial adoption.

Enzymatic hydrolysis, another green method, involves cellulolytic enzymes digesting cellulose substrates. Though traditionally designed to maximize fermentable sugars from cellulose pulps, this method can also yield nanocellulose particles (Squinca et al., 2020; Bondancia et al., 2017). The process involves endoglucanases randomly attacking internal β-1,4-glycosidic bonds, exoglucanases progressively cleaving cellobiose units, and β-glucosidases hydrolyzing cellobiose into glucose monomers, resulting in cellulose nanocrystals (CNC) and cellulose nanofibrils (CNF) (Ruiz et al., 2012). However, the high costs of acquiring appropriate enzymes and low nanocellulose yields present significant challenges to large-scale application.

Recently, DES and NADES treatments have emerged as promising alternatives for hydrolyzing cellulose fibers (Almeida et al., 2023). These solvents, formulated with organic components, are considered natural and non-toxic, but their environmental interactions require careful evaluation (Martínez et al., 2022). Notably, NADES have demonstrated recycling potential in processing lignocellulosic biomass, allowing the recovery of components from eutectic systems and reducing processing costs (Almeida et al., 2023).

The exploration of these green technologies, especially NADES treatments, as alternatives to conventional acid hydrolysis for producing nanocellulose is gaining momentum. These innovative methods offer the potential to enhance production efficiency, reduce environmental impact, and address the limitations of traditional processes. However, optimizing the high energy demands remains crucial for achieving sustainable and scalable nanocellulose production.

9.5 DES and NADES in Nanocellulose Extraction

DES offer highlightable physicochemical properties favorable for the extraction of compounds of interest. The synthesis of DES was first reported by Abbott et al. in 2001, with the initial DES comprising choline chloride (ChCl) and urea (Abbott et al., 2004; Abranches & Coutinho, 2023; Benvenutti et al., 2019; Fernández et al., 2018).

As stated before, DES have gathered attention as alternative and sustainable solvents due to their physical properties and lower toxicity. When DES are derived from naturally occurring metabolites (plants and/or animals), they are referred to as natural deep eutectic solvents (NADES) (Abranches & Coutinho, 2023; Carreón-Hidalgo et al., 2024).

9.5.1 Chemical Composition

There are four recognized types of Deep Eutectic Solvents (DES): Type I, a quaternary ammonium salt combined with a metal halide or imidazole salt; Type II, formulated with a quaternary ammonium salt and a hydrated metal halide, such as Choline Chloride (ChCl); Type III, composed of a hydrogen bond donor (HBD) and a hydrogen bond acceptor (HBA), and is the most commonly used type; and Type IV, consisting of a metal halide and urea (Santana-Mayor et al., 2021).

Quaternary ammonium salts (like ChCl) act as HBAs, while amino acids (e.g., alanine, glycine betaine), organic acids (e.g., malic acid, oxalic acid), and carbohydrates (e.g., fructose, glucose) serve as HBDs (Almeida et al., 2023; Benvenutti et al., 2019). Heating these components generates intramolecular interactions, such as hydrogen bonding, Van der Waals forces, and electrostatic forces, and lead to complex structural networks (Fig. 9.3).

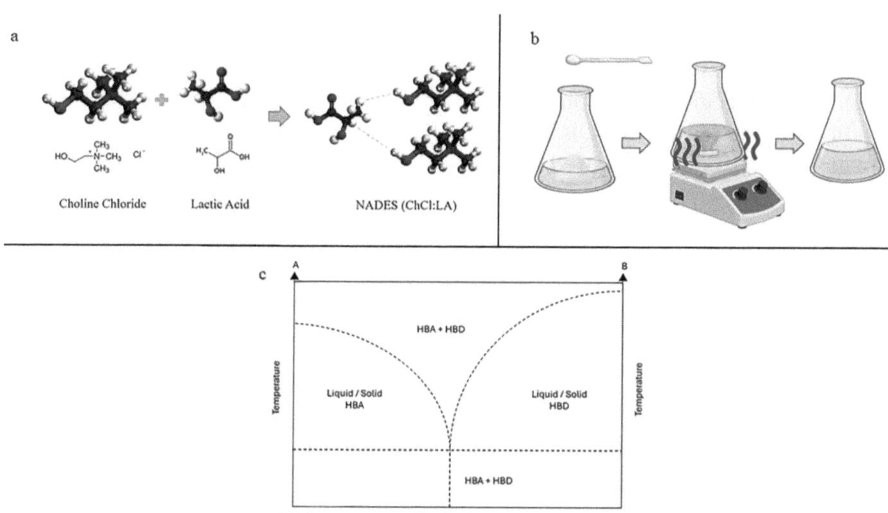

Fig. 9.3 NADES preparation process (ChCl:LA). (**a**) Chemical process; (**b**) preparation in laboratory; and (**c**) NADES phases diagram

Table 9.1 Examples of different HBA and HBD for NADES formation

HBA	HBD	Molar ratio	Temperature	Ref
ChCl	Glycerol	1:2 M	40 °C	García et al. (2016)
	Lactic acid	1:2 M		
	Urea	1:2 M		
	Sucrose	1:1 M / 4:1 M		
	1,4-Butanediol	1:5 M		
	Xylitol	2:1 M		
	1,2-Propanediol	1:1 M		
	Malonic acid	1:1 M		
	Urea/glycerol	1:1:1 M		
D-Fructose	D-glucose/ sucrose	1:1:1 M		
ChCl	Lactic acid	1:4 M	80 °C	Jiménez-Ortega et al. (2023)
	1,2-Butanediol	1:2 M		
ChCl	Acetic acid	1:2 M / 2:1 M	–	Lomba et al. (2021)
	Lactic acid	1:2 M	–	
	Citric acid	1:2 M	–	
	Glycolic acid	1:2 / 2:1 M	–	
ChCl	D + Glucose	1:1 M	40 °C	Paiva et al. (2014)
	L+ tartaric acid	2:1 M	40 °C	
	D-xylose	2:1 M / 3:1 M	40 °C	
D (+) glucose	L+ tartaric acid	1:1 M	–	
ChCl	Oxalic acid	1:1 M	60 °C	Ruiz Deance et al. (2020)

Different DES/NADES compositions are prepared at various molar ratios and conditions. ChCl is the most used HBA, while HBDs like urea, lactic acid, glucose, and glycerol are commonly used (Santana-Mayor et al., 2021) as can be observed on Table 9.1.

9.5.2 Physicochemical Properties

DES and NADES are distinguished by their ability to achieve a significantly lower melting point when a hydrogen bond acceptor (HBA) and a hydrogen bond donor (HBD) are mixed in specific molar ratios. This phenomenon, known as the eutectic point, is crucial as it allows DES and NADES to remain liquid at room temperature, offering a considerable advantage for chemical and biotechnological processes by facilitating efficient compound extraction (Abranches & Coutinho, 2023; Carreón-Hidalgo et al., 2024; Benvenutti et al., 2019; Boateng, 2023).

In addition to lowering the melting point, these interactions also impact other physicochemical properties such as viscosity, density, pH, and polarity, which are directly influenced by the chemical composition of the eutectic mixture. These properties play a significant role in the extraction process of compounds of interest (Contreras-Hernández et al., 2018; Fernández et al., 2018).

Viscosity, in particular, is a critical property of DES/NADES, varying significantly depending on the nature of the individual components and affecting the fluidity and efficiency of mass transfer in chemical processes, directly influencing the dynamics and efficiency of these processes. When viscosity is high, it can be lowered by adding water, which enhances the operability of DES in various applications and optimizes extraction and reaction processes (Almeida et al., 2023; Fernández et al., 2018).

Density is another fundamental property, influenced by the structure of the DES, their molar ratios, and the interactions between their constituents. The density of DES decreases with increasing temperature, which is an important consideration in their application (Contreras-Hernández et al., 2018).

pH is also a critical aspect of DES, as the relative acidity of the cationic and anionic species in the mixture can significantly influence the pH, affecting viscosity, density, and the efficiency of compound extraction processes (Boateng, 2023).

Finally, DES generally exhibit hydrophilic characteristics due to their high electronegativity and ability to form hydrogen bonds. However, some eutectic mixtures can display both polar and nonpolar properties, expanding their application range and enabling their use in various processes requiring versatility in solubility and molecular interactions (Benvenutti et al., 2019).

These properties are of utmost importance in DES, since the constituents of each mixture directly affect the transport phenomenon involved in the extraction process (Benvenutti et al., 2019). Therefore, understanding how the structure and molar composition of DES influence their physical properties and how these, in turn, impact extraction processes is essential to optimize their use in industrial and scientific applications.

9.5.3 Action Mechanisms on Nanocellulose Extraction

Considering that acid hydrolysis is the conventional method for obtaining nanocellulose, there is a growing interest in exploring processes that preserve the desirable properties of nanocellulose while employing environmentally friendly methods.

Both acid hydrolysis and DES nanocellulose production methods rely on solvents to infiltrate the glycosidic bonds within the amorphous regions of cellulose, which are inherently disordered and thus more susceptible to disruption. Conversely, these solvents encounter greater difficulty in penetrating the crystalline regions due to their more organized structure, ultimately resulting in the production of nanocellulose (Almeida et al., 2023).

9.6 Nanocellulose Obtention in Latin America

Lignocellulosic waste represents a significant source of waste worldwide, with an approximate production of 2.3 billion tons, with Brazil being the first country in Latin America to generate this type of waste, followed by Argentina and Mexico (FAO, 2023). This waste contains components that have excellent potential to produce new products, creating a new area of opportunity and generating value-added products. In this context, the Latin American region is positioned as an area of strategic interest for research and development in the extraction of compounds such as nanocellulose.

In this section, several case studies are presented that highlight the efforts and advances in the production of nanocellulose in different Latin American countries. These examples allow us to analyze the methods used and the diversity of sources used for the extraction of nanocellulose, providing a detailed overview of the current state of technology and practices implemented in these countries, underlining the importance of innovation and sustainable development to promote competitive progress in the field of nanocellulose.

Some studies carried out on the production of nanocellulose with different green technologies in Latin America, where different agro-industrial wastes were used, considered lignocellulosic waste, are presented in Table 9.2.

Ruiz Deance et al. (2020) explored the CNC production by treating mercerized cotton with a NADES composed of ChCl and oxalic acid dihydrate. Their process, which spanned up to 60 days at temperatures ranging from room temperature to 100 °C, resulted in CNC particles sized 50–1500 nm, which were subsequently used in synthesizing macroporous polymers.

Building on similar pursuits in nanocellulose extraction, González-Soto et al. (2016), in Mexico, utilized agro-industrial banana residues, applying autoclaving at 150 Pa for 1.5 hours at 125 °C to obtain CNC. Further advancing the field, Cherian et al. (2010), in Brazil, employed steam explosion at 20 lb. pressure for 1 hour, successfully deriving CNF from pineapple leaves. This steam explosion method was also effectively used by Ospina-Varón et al. (2022), in Colombia, to produce CNC from sugarcane bagasse. Meanwhile, Furtado et al. (2021) in Brazil achieved CNF from bleached pulp using a grinding process with a 150 mm diameter at 1500 rpm.

Complementing these mechanical and thermal methods, enzymatic hydrolysis has been studied as a significant approach for nanocellulose extraction. Hernández Pérez et al. (2023), in Mexico, demonstrated this by using rice husk waste along with a combination of enzymes, including D-(+)-cellobiose and endo-1,4-β-D-glucanase, to produce CNC. Similarly, Bondancia et al. (2017) in Brazil obtained CNC from eucalyptus pulp with an 83% crystallinity index, a length of 260 nm, and a diameter of 15 nm. Further expanding on this enzymatic approach, Reyes-González (2023), in Chile, produced CNF from bleached eucalyptus kraft pulp. In parallel, Hernández-Varela et al. (2021), in Mexico, utilized ball milling to derive CNC from garlic skin and agave leaves, showcasing an alternative mechanical method.

Table 9.2 Nanocellulose obtained from diverse lignocellulosic biomasses by green technologies in Latin America

Green technology	Biomass source	Country of Latin America	Type of nanocellulose	Authors
NADES (ChCl: OAD) (1:1 M)	Mercerized cotton	Mexico	CNC	Ruiz Deance et al. (2020).
Steam explosion	Banana peel, pseudostem and rachis	Mexico	NC	González-Soto et al. (2016)
	Pineapple leaves	Brazil	CNF	Cherian et al. (2010).
	Sugarcane bagasse	Colombia	NC	Ospina-Varón et al. (2022)
Grinding	Bleached pulp	Brazil	CNF	Furtado et al. (2021)
Enzymatic hydrolysis	Rice husk waste	Mexico	NC	Hernández Pérez et al. (2023)
	Eucalyptus pulp	Brazil	CNC	Bondancia et al. (2017)
	Bleached Eucalyptus Kraft Pulp	Chile	CNF	Reyes-González (2023)
Ball milling	Garlic skin and *Agave* leaves	Mexico	CNP	Hernández-Varela et al. (2021)
High pressure homogenization	Bleached *Agave tequilana* leaves	Mexico	CNF	Hernández-Francisco et al. (2020).
	Bleached *Agave tequilana* bagasse	Mexico	CNF	Palacios Hinestroza et al. (2019)
	Acetylated *Agave tequilana* bagasse	Mexico	CNF	Sulbarán-Rangel et al. (2020)
	Bleached *Agave tequilana* leaves and bagasse	Mexico	CNF	Robles et al. (2018)
Mechanical treatment	Bell pepper (*Capsicum annuum*), tomato (*Solanum lycopersicum*), and eggplant (*Solanum melongena*) horticultural residues	Spain	CNF	Haro-Niza et al. (2022)
Ultrasound enhanced acid hydrolysis	Leaves, raw pineapple and agave bagasse	Mexico	NC	Vidales-Carrillo (2019)
	Bleached *Agave tequilana* leaves and bagasse	Mexico	CNC	Robles et al. (2018).
	Microcrystalline powder	Brazil	CNC	Meirelles et al. (2020)

CNP cellulose nanoparticles; *CN* nanocellulose

High-pressure homogenization stands out as one of the most studied green technologies for nanocellulose production in Ibero-America. In Mexico, Hernández-Francisco et al. (2020) obtained CNF from bleached *Agave tequilana* leaves, while Palacios Hinestroza et al. (2019) used a microfluidizer with 400 and 200 μm chambers on bleached *Agave tequilana* bagasse to produce CNF. Sulbarán-Rangel et al. (2020) expanded on these findings by extracting CNF from acetylated *Agave tequilana* bagasse.

In Latin American region, mechanical treatment was also employed by Haro-Niza et al. (2022) in Spain, where CNF was obtained from bell pepper, tomato, and eggplant horticultural residues.

Finally, ultrasound technology has proven effective in nanocellulose production by different Latin American authors. Vidales-Carrillo (2019) used sonification with acid hydrolysis on raw pineapple, agave bagasse, and leaves not used in pulque production to obtain nanocellulose. Similarly, Robles et al. (2018) in Mexico produced nanocellulose from blue agave leaves and bagasse fibers using sonochemical acid hydrolysis with sulfuric acid, yielding 40% for bagasse and 51% for leaves, with a width of 33 ± 13 nm and a length of 350 ± 153 nm. Additionally, Meirelles et al. (2020) in Brazil used hydrochloric acid hydrolysis followed by ultrasonication to obtain nanocellulose for Pickering emulsions.

9.7 Challenges and New Directions in Nanocellulose Extraction with NADES

The main producing countries of lignocellulosic waste are in the Latin American region, presenting a significant opportunity to add value to this waste, thereby generating a circular bioeconomy and the recovery of lignocellulosic waste, capable of producing products such as nanocellulose. Nanocellulose has various applications, including its use as a stabilizer of emulsions, reinforcement in biopolymeric films, and in the production of cosmetic and personal hygiene products, to mention a few.

However, there is currently very scarce literature from Latin America on the extraction of nanocellulose using green technologies, and traditional methods, such as acid hydrolysis, are highly polluting. Nevertheless, the review of the literature indicates that in Latin America there is a wide field of opportunities for the extraction of nanocellulose using green methods, specifically with DES and NADES, which are environmentally friendly. These technologies can enhance the extraction of nanocellulose, leveraging lignocellulosic waste from the region and promoting sustainability and innovation in this field.

References

Abbott, A. P., Boothby, D., Capper, G., Davies, D. L., & Rasheed, R. K. (2004). Deep eutectic solvents formed between choline chloride and carboxylic acids: versatile alternatives to ionic liquids. *Journal of the American Chemical Society, 126*(29), 9142–9147. https://doi.org/10.1021/ja048266j

Abranches, D. O., & Coutinho, J. A. P. (2023). Everything you wanted to know about deep eutectic solvents but were afraid to be told. *Annual Review of Chemical and Biomolecular Engineering, 14*, 141–163. https://doi.org/10.1146/annurev-chembioeng-101121-085323

Almeida, R. O., Maloney, T. C., & Gamelas, J. A. (2023). Production of functionalized nanocelluloses from different sources using deep eutectic solvents and their applications. *Industrial Crops and Products, 199*, 116583. https://doi.org/10.1016/j.indcrop.2023.116583

Anastas, P. T., & Warner, J. C. (1998). Principles of green chemistry. *Green Chemistry: Theory and Practice, 29*, 14821–14842.

Arevalo-Gallegos, A., Ahmad, Z., Asgher, M., Parra-Saldivar, R., & Iqbal, H. M. N. (2017). Lignocellulose: A sustainable material to produce value-added products with a zero waste approach—A review. *International Journal of Biological Macromolecules, 99*, 308–318. https://doi.org/10.1016/j.ijbiomac.2017.02.097

Benvenutti, L., Zielinski, A. A. F., & Ferreira, S. R. S. (2019). Which is the best food emerging solvent: IL, DES or NADES? *Trends in Food Science & Technology, 90*, 133–146. https://doi.org/10.1016/j.tifs.2019.06.003

Boateng, I. D. (2023). Evaluating the status quo of deep eutectic solvent in food chemistry. Potentials and limitations. *Food Chemistry, 406*, 135079. https://doi.org/10.1016/j.foodchem.2022.135079

Bondancia, T. J., Mattoso, L. H. C., Marconcini, J. M., & Farinas, C. S. (2017). A new approach to obtain cellulose nanocrystals and ethanol from eucalyptus cellulose pulp via the biochemical pathway. *Biotechnology Progress, 33*(4), 1085–1095. https://doi.org/10.1002/btpr.2486

Carreón-Hidalgo, J. P., de Lourdes Ruiz-Peralta, M., & Corona-Jiménez, E. (2024). Solventes eutécticos profundos naturales: Aplicaciones en biotecnología y otros sectores industriales Natural deep eutectic solvents. *Applications in biotechnology and other industrial sectors., 28*(2).

Chen, H., Liu, J., Chang, X., Chen, D., Xue, Y., Liu, P., Lin, H., & Han, S. (2017). A review on the pretreatment of lignocellulose for high-value chemicals. *Fuel Processing Technology, 160*, 196–206. https://doi.org/10.1016/j.fuproc.2016.12.007

Cherian, B. M., Leão, A. L., de Souza, S. F., Thomas, S., Pothan, L. A., & Kottaisamy, M. (2010). Isolation of nanocellulose from pineapple leaf fibres by steam explosion. *Carbohydrate Polymers, 81*(3), 720–725. https://doi.org/10.1016/j.carbpol.2010.03.046

Contreras-Hernández, M. G., Ochoa-Martínez, L. A., Rutiaga-Quiñones, J. G., Rocha-Guzmán, N. E., Lara-Ceniceros, T. E., Contreras-Esquivel, J. C., Prado Barragán, L. A., & Rutiaga-Quiñones, O. M. (2018). Effect of ultrasound pre-treatment on the physicochemical composition of Agave durangensis leaves and potential enzyme production. *Bioresource Technology, 249*, 439–446. https://doi.org/10.1016/j.biortech.2017.10.009

Devi, A., Bajar, S., Kour, H., Kothari, R., Pant, D., & Singh, A. (2022). Lignocellulosic biomass valorization for bioethanol production: a circular bioeconomy approach. *Bioenergy Research, 15*(4), 1820–1841. https://doi.org/10.1007/s12155-022-10401-9

FAO. (2023). Food and Agriculture Organization of the United Nations.

Fernández, M. d. l. Á., Boiteux, J., Espino, M., Gomez, F. J. V., & Silva, M. F. (2018). Natural deep eutectic solvents-mediated extractions: The way forward for sustainable analytical developments. *Analytica Chimica Acta, 1038*, 1–10. https://doi.org/10.1016/j.aca.2018.07.059

Furtado, M. R., da Matta, V. M., Carvalho, C. W. P., Magalhães, W. L. E., Rossi, A. L., & Tonon, R. V. (2021). Characterization of spray-dried nanofibrillated cellulose and effect of different homogenization methods on the stability and rheological properties of the reconstituted suspension. *Cellulose, 28*, 207–221. https://doi.org/10.1007/s10570-020-03516-9

García, A., Gandini, A., Labidi, J., Belgacem, N., & Bras, J. (2016). Industrial and crop wastes: A new source for nanocellulose biorefinery. *Industrial Crops and Products, 93*, 26–38. https://doi.org/10.1016/j.indcrop.2016.06.004

George, J., & Sabapathi, S. N. (2015). Cellulose nanocrystals: synthesis, functional properties, and applications. *Nanotechnology, Science and Applications, 8*, 45–54. https://doi.org/10.2147/NSA.S64386

González-Soto, R. A., Reyes-Atrizco, J. N., Gutiérrez-Meráz, F., & Pacheco-Vargas, G. (2016). Nanocelulosa obtenida de residuos agroindustriales del cultivo de plátano macho (*Musa paradisiaca* L.). *Investigación y Desarrollo en Ciencia y Tecnología de Alimentos, 1*(2), 301–306.

Haro-Niza, J., Rincón, E., Gonzalez, Z., Espinosa, E., & Rodríguez, A. (2022). Nanocellulose from Spanish harvesting residues to improve the sustainability and functionality of linerboard recycling processes. *Nanomaterials, 12*(24), 4447. https://doi.org/10.3390/nano12244447

Heinze, T. (2016). Cellulose: structure and properties. In *Cellulose chemistry and properties: fibers, nanocelluloses and advanced materials* (pp. 1–52).

Hernández Pérez, R., Olarte Paredes, A., Salgado Delgado, R., Salgado Delgado, A. M., Medrano, A. S., & Martínez Candia, F. R. (2023). Manejo de la cascarilla de arroz como residuo postcosecha y su conversión en nanocelulosa. *Mundo nano. Revista interdisciplinaria en nanociencias y nanotecnología, 16*(30), 10.22201/ceiich.24485691e.2023.30.69697.

Hernández-Francisco, E., Bonilla-Cruz, J., Márquez-Lamas, U., Suárez-Jacobo, Á., Longoria-Rodríguez, F., Rivera-Haro, J., Russell, P., Ali, Z., Yin, C. Y., & Lara-Ceniceros, T. E. (2020). Entangled cellulose nanofibrils/nanosheets derived from native mexican agave for lead(II) ion removal. *Cellulose, 27*(15), 8785–8798. https://doi.org/10.1007/s10570-020-03373-6

Hernández-Varela, J. D., Chanona-Pérez, J. J., Calderón Benavides, H. A., Cervantes Sodi, F., & Vicente-Flores, M. (2021). Effect of ball milling on cellulose nanoparticles structure obtained from garlic and agave waste. *Carbohydrate Polymers, 255*. https://doi.org/10.1016/j.carbpol.2020.117347

Horvath, I. T., & Anastas, P. T. (2007). Innovations and green chemistry. *Chemical Reviews, 107*(6), 2169–2173. https://doi.org/10.1021/cr078380v

Huang, J., Ma, X., Yang, G., & Alain, D. (2019). Introduction to nanocellulose. In *Nanocellulose: From fundamentals to advanced materials* (pp. 1–20). https://doi.org/10.1002/9783527807437.ch1

Jiang, J., Zhu, Y., & Jiang, F. (2021). Sustainable isolation of nanocellulose from cellulose and lignocellulosic feedstocks: Recent progress and perspectives. *Carbohydrate Polymers, 267*, 118188. https://doi.org/10.1016/j.carbpol.2021.118188

Jiménez-Ortega, L. A., Valdez-Baro, O., Heredia-Bátiz, J. M., García-Estrada, R. S., & Heredia, J. B. (2023). Control de fitopatógenos con extractos de biomasa de chile y de maíz. *Tip Revista Especializada en Ciencias Químico-Biológicas, 26*. https://doi.org/10.22201/fesz.23958723e.2023.576

Lomba, L., Ribate, M. P., Sangüesa, E., Concha, J., Garralaga, M. ª. P., Errazquin, D., García, C. B., & Giner, B. (2021). Deep eutectic solvents: Are they safe? *Applied Sciences (Basel), 11*(21), 10061. https://doi.org/10.3390/app112110061

Martínez, G. M., Townley, G. G., & Martínez-Espinosa, R. M. (2022). Controversy on the toxic nature of deep eutectic solvents and their potential contribution to environmental pollution. *Heliyon, 8*(12). https://doi.org/10.1016/j.heliyon.2022.e12567

Matiacevich, S., Soto-Madrid, D., & Gutiérrez-Cutiño, M. (2023). Economía circular: obtención y encapsulación de compuestos polifenólicos provenientes de residuos agroindustriales. *Rivar, 10*(28), 77–100. https://doi.org/10.35588/rivar.v10i28.5343

Meirelles, A. A. D., Costa, A. L. R., & Cunha, R. L. (2020). Cellulose nanocrystals from ultrasound process stabilizing O/W Pickering emulsion. *International Journal of Biological Macromolecules, 158*, 75–84. https://doi.org/10.1016/j.ijbiomac.2020.04.185

Morehead, F. F. (1950). Ultrasonic disintegration of cellulose fibers before and after acid hydrolysis. *Textile Research Journal, 20*(8), 549–553. https://doi.org/10.1177/004051755002000803

Murcia Fandiño, J. S., & Esquiaqui Marín, L. A. (2021). Química verde aplicada en los residuos de universidades. *Educación Química, 32*(2), 154–167. https://doi.org/10.22201/fq.18708404e.2021.2.76534

Nechyporchuk, O., Belgacem, M. N., & Bras, J. (2016). Production of cellulose nanofibrils: A review of recent advances. *Industrial Crops and Products, 93*, 2–25. https://doi.org/10.1016/j.indcrop.2016.02.016

Ojo, A. (2023). An overview of lignocellulose and its biotechnological importance in high-value product production. *Fermentation, 9*(11). https://doi.org/10.3390/fermentation9110990

Ospina-Varón, R.A., López-Suárez, F.E., & Aristizábal-Marulanda, V. (2022). Prefeasibility study for the nanocellulose production from biomass in the Colombian context. Biomass Conversion and Biorefinery, 10, 1–12. https://doi.org/10.1007/s13399-021-02076-7.

Paiva, A., Craveiro, R., Aroso, I., Martins, M., Reis, R. L., & Duarte, A. R. C. (2014). Natural deep eutectic solvents–solvents for the 21st century. *ACS Sustainable Chemistry & Engineering, 2*(5), 1063–1071. https://doi.org/10.1021/sc500096j

Palacios Hinestroza, H., Hernández Diaz, J. A., Esquivel Alfaro, M., Toriz, G., Rojas, O. J., & Sulbarán-Rangel, B. C. (2019). Isolation and characterization of nanofibrillar cellulose from Agave tequilana Weber bagasse. *Advances in Materials Science and Engineering, 2019*(1), 1342547. https://doi.org/10.1155/2019/1342547

Peñaranda-González, L. V., Montenegro-Gómez, S. P., & Giraldo-Abad, P. (2017). Exploitation of agroindustrial waste in Colombia. *Revista De Investigación Agraria Y Ambiental, 8*(2).

Pereira, B., & Arantes, V. (2018). Nanocelluloses from sugarcane biomass. In *Advances in sugarcane biorefinery* (pp. 179–196). Elsevier. https://doi.org/10.1016/B978-0-12-804534-3.00009-4

Phanthong, P., Reubroycharoen, P., Hao, X., Xu, G., Abudula, A., & Guan, G. (2018). Nanocellulose: Extraction and application. *Carbon Resources Conversion, 1*(1), 32–43. https://doi.org/10.1016/j.crcon.2018.05.004

Pino, A. L. (2020). *Química verde: Enfoque sistémico.* Ediciones UNL. https://bibliotecavirtual.unl.edu.ar/handle/11185/5630.

Reyes-González, I. A. (2023). Pretratamiento químico-enzimático de pulpa Kraft de Eucalipto para la producción de nanocelulosa fosforilada.

Robles, E., Fernández-Rodríguez, J., Barbosa, A. M., Gordobil, O., Carreño, N. L. V., & Labidi, J. (2018). Production of cellulose nanoparticles from blue agave waste treated with environmentally friendly processes. *Carbohydrate Polymers, 183*, 294–302. https://doi.org/10.1016/j.carbpol.2018.01.015

Roy, R., Rahman, M. S., & Raynie, D. E. (2020). Recent advances of greener pretreatment technologies of lignocellulose. *Current Research in Green and Sustainable Chemistry, 3.* https://doi.org/10.1016/j.crgsc.2020.100035

Ruiz Deance, A. L., Mota Morales, J. D., Universidad, N. A. de M., & Universidad, N. A. de M. (2020). Obtención de nanocristales de celulosa mediante líquidos eutécticos usando conceptos de química verde para la síntesis de polímeros macroporosos.

Ruiz, H. A., Vicente, A. A., & Teixeira, J. A. (2012). Kinetic modeling of enzymatic saccharification using wheat straw pretreated under autohydrolysis and organosolv process. *Industrial Crops and Products, 36*(1), 100–107. https://doi.org/10.1016/j.indcrop.2011.08.014

Santana-Mayor, Á., Rodríguez-Ramos, R., Herrera-Herrera, A. V., Socas-Rodríguez, B., & Rodríguez-Delgado, M. Á. (2021). Deep eutectic solvents. The new generation of green solvents in ytical chemistry. *Trends in Analytical Chemistry, 134*, 116108. https://doi.org/10.1016/j.trac.2020.116108

Santos, T. M., Souza Filho, M. d. S. M., Caceres, C. A., Rosa, M. F., Morais, J. P. S., Pinto, A. M. B., & Azeredo, H. M. C. (2014). Fish gelatin films as affected by cellulose whiskers and sonication. *Food Hydrocolloids, 41*, 113–118. https://doi.org/10.1016/j.foodhyd.2014.04.001

Scapini, T., Dos Santos, M. S. N., Bonatto, C., Wancura, J. H. C., Mulinari, J., Camargo, A. F., Klanovicz, N., Zabot, G. L., Tres, M. V., Fongaro, G., & Treichel, H. (2021). Hydrothermal pretreatment of lignocellulosic biomass for hemicellulose recovery. *Bioresource Technology, 342*, 126033. https://doi.org/10.1016/j.biortech.2021.126033

Squinca, P., Bilatto, S., Badino, A. C., & Farinas, C. S. (2020). Nanocellulose production in future biorefineries: An integrated approach using tailor-made enzymes. *ACS Sustainable Chemistry and Engineering, 8*(5), 2277–2286. https://doi.org/10.1021/acssuschemeng.9b06790

Sulbarán-Rangel, B., Hernández Díaz, J. A., Guzmán González, C. A., & Rojas, O. J. (2020). Partially acetylated cellulose nanofibrils from Agave tequilana bagasse and Pickering stabilization. *Journal of Dispersion Science and Technology, 43*(9), 1391–1398. https://doi.org/10.1080/01932691.2020.1858855

Vidales-Carrillo, H. (2019). *Métodos de Extracción de Nanocelulosa a partir de Residuos de Agave spp*. Universidad Tecnológica del Valle de Toluca.

Wulandari, W. T., Rochliadi, A., & Arcana, I. M. (2016). Nanocellulose prepared by acid hydrolysis of isolated cellulose from sugarcane bagasse. In *IOP conference series: materials science and engineering, 107*(1), p. 012045. IOP Publishing. https://doi.org/10.1088/1757-899X/107/1/012045.

Chapter 10
Analysis of Agroecological Bioeconomy

Carlos Alberto Zuniga González, Silvio Javier Sáenz Rojas, Pedro José Toruño, and Walter Alberto Reyes Canales

Abstract Background: In response to the global economic, social, and environmental crises, Latin America has been driving the advancement of the bioeconomy with a commitment to substitute fossil fuels with "renewable resources." However, pressing questions arise regarding the backdrop of this proposal. Who stands to gain, and who might lose? Additionally, the agroecological approach raises fundamental inquiries about its usage and interventionist role in ecosystems and resources. Methodology: This paper investigates the scientific literature on issues related to agricultural bioeconomy by applying a bibliometric network analysis. Bibliometric analysis is applied to the publications of the Scopus database during the period 2010–2023 in order to provide an overview of the main aspects that characterize agricultural bioeconomy. Results: An "agroecological bioeconomic" perspective values the alternative of those who argue that agro industrial monocultures make farmers dependent on agrochemicals. Conclusions: External contributions undermine and distance the consumer's knowledge base. In Latin America, defenders of this point of view come from the public and private sectors. Research institutes on

C. A. Zuniga González (✉)
Research Center on Bioeconomy and Climate Change, Specific Direction of Agroecology, Direction of Agricultural and Veterinary Sciences, National Autonomous University of Nicaragua, Leon, Nicaragua
e-mail: czuniga@ct.unanleon.edu.ni

S. J. Sáenz Rojas · W. A. Reyes Canales
Direction of Agricultural and Veterinary Sciences, Specific Direction of Agroecology, National Autonomous University of Nicaragua, Leon, Nicaragua
e-mail: silvio.saenz@ev.unanleon.edu.ni; walter.reyes@ev.unanleon.edu.ni

P. J. Toruño
Bio-refinery-Bio-products Research Unit, Research Center on Bioeconomy and Climate Change, Faculty of Agricultural and Veterinary Sciences, Specific Direction of Agroecology, National Autonomous University of Nicaragua, Leon, Nicaragua
e-mail: pjoseto@ev.unanleon.edu.ni

S. G. Ceballos Pérez et al. (eds.), *Sustainability in Latin America*,
https://doi.org/10.1007/978-3-032-09787-3_10

organic farming and movements to support them including the organic department of specialized agricultural organizations and environmental NGOs. Therefore, this bibliographic review is motivated toward an agroecological bioeconomy.

Keywords Biomass · Bio inputs · Bio-based economy · Innovation · Biotechnology

10.1 Introduction

The concept of agroecological bioeconomy represents a multidisciplinary field that amalgamates elements from agroecology and bioeconomy. Agroecology emphasizes sustainable agricultural practices, considering ecological, economic, and social aspects, while bioeconomy aims to substitute fossil-based resources with renewable alternatives. The convergence of these two concepts creates avenues for industries and agriculture to tackle resource limitations and encourage sustainable land and resource utilization (D'Amato et al., 2020; Pimbert, 2018; Meynard, 2017; Tomich et al., 2011). Despite the promise of a sustainable bioeconomy, it encounters intricate challenges as it intertwines with broader social relations, potentially perpetuating global inequalities in trade and power dynamics. Despite the increasing prevalence of agroecology, its integration into Canadian food policy and practice remains constrained (Backhouse et al., 2022; Isaac et al., 2018). Overall, the agroecological bioeconomy offers a potential pathway for societal transformation toward sustainability. However, it necessitates careful consideration of social, economic, and environmental factors. The challenge lies in navigating these complex intersections to strike a balance that drives sustainability both globally and locally (Lima, 2022; von Braun, 2018; Giraldo & Rosset, 2018; Foguesatto et al., 2017).

There is a significant gap in integrating agroecological principles into broader agricultural and economic policies. Bridging this gap is crucial for ensuring that the principles of agroecology are not just theoretical but are practically implemented within regulatory frameworks (Wezel et al., 2020).

While there is growing interest in agroecology, there is a gap in effectively translating this knowledge into actionable strategies, especially within different socio-economic contexts and geographic regions. This translation is vital for widespread adoption and adaptation (Feleke et al., 2021).

Understanding how agroecological practices can be scaled up without losing their context-specific effectiveness is a critical research gap. What works on a small-scale might not necessarily translate seamlessly to larger agricultural systems, and this discrepancy needs further investigation (Schwarz et al., 2022).

There is a need for comprehensive studies assessing the socioeconomic impacts of transitioning toward an agroecological bioeconomy. This includes understanding its effects on livelihoods, income distribution, and community resilience (Gil et al., 2022). Research could delve into technological innovations that support and enhance agroecological practices. This might include exploring precision agriculture technologies tailored to agroecological principles or finding ways to integrate digital tools effectively into these systems (Feleke et al., 2021).

Investigating consumer behavior and market dynamics concerning agroecological products is crucial. Understanding consumer preferences, willingness to pay, and market incentives for these products can drive the adoption of sustainable agricultural practices (Pyka et al., 2022).

While agroecology focuses on sustainability, there is a gap in long-term studies that comprehensively assess its environmental impacts, particularly concerning soil health, biodiversity conservation, and resilience to climate change over extended periods (Von Cossel et al., 2019).

Filling these gaps in research can provide a more holistic understanding of agroecological bioeconomy, facilitating its wider adoption and integration into global agricultural systems.

Furthermore, research gaps necessitate a comprehensive understanding of the current state of knowledge and research trends in agroecological bioeconomy. A bibliometric study focused on agroecological bioeconomy holds paramount importance in this context.

Such a study serves as a crucial foundation for advancing research in this field. By analyzing and synthesizing existing literature, a bibliometric study provides a panoramic view of research patterns, identifying key thematic areas, influential authors, prolific journals, and emerging trends within agroecological bioeconomy. This study acts as a compass, guiding researchers and policymakers toward critical areas that require further investigation and intervention. It helps in pinpointing specific gaps in knowledge translation, policy integration, technological innovations, socioeconomic impacts, environmental sustainability, and consumer behavior related to agroecological bioeconomy.

Moreover, a bibliometric analysis aids in establishing a baseline understanding, enabling future research to build upon existing knowledge, avoid redundancy, and foster interdisciplinary collaboration. It facilitates the identification of opportunities for scaling up successful practices while considering context-specific challenges. Ultimately, this study serves as a strategic tool for decision-makers, stakeholders, and researchers, offering insights that can shape policies, guide funding priorities, and steer research efforts toward fostering a more sustainable, equitable, and resilient agroecological bioeconomy.

10.2 Literature Review

The concept of agroecological bioeconomy (Fig. 10.1) encompasses the sustainable and innovative use of biomass and biological knowledge to provide food, feed, industrial products, bioenergy, and ecological services (Lewandowski, 2018; Meadows et al., 1972). It involves an interdisciplinary approach, integrating natural sciences, economic and social sciences, politics, and action (Meynard, 2017).

Agroecology, as a component of the agroecological bioeconomy, focuses on understanding and working with the interactions between plants, animals, humans, and the environment within agricultural systems (Ryan & Peigné, 2017). The

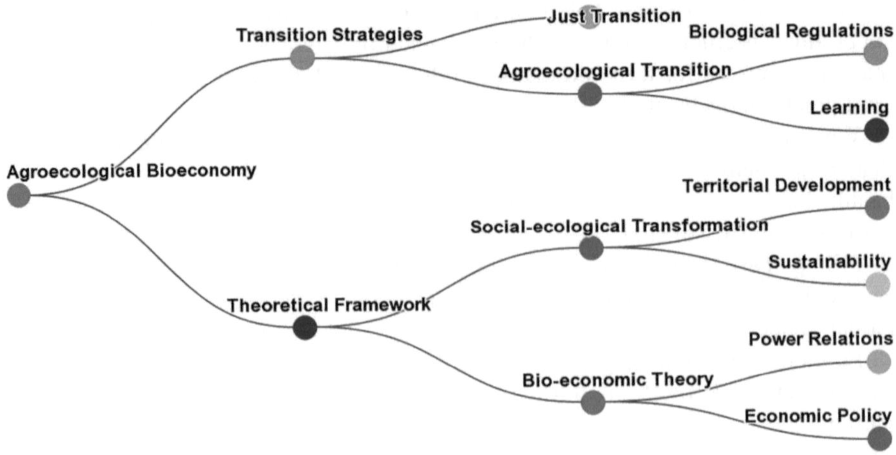

Fig. 10.1 The concept of agroecological bioeconomy (Lewandowski, 2018; Meadows et al., 1972)

bioeconomy trend is driven by the need to address resource constraints related to climate, water, energy, and land, as well as advances in microbiology and shifts in consumer preferences (von Braun, 2018). However, there is still no consensus on the grammatical structure of the term "bioeconomy" (Viaggi, 2018).

The bioeconomy is also a contested field, with alternative, more socio-ecological and locally embedded approaches being underrepresented in the debate (Von Braun, 2018). Overall, the agroecological bioeconomy aims to provide sustainable alternatives to the fossil-based economy and promote resource protection and food security (Von Braun, 2018).

Agroecological bioeconomy is a concept that combines principles and practices from agroecology and bioeconomy. Agroecological bioeconomy is based on a set of principles that emphasize sustainable production, resilience of agroecosystems, holistic approaches, and scaling up actions at different levels (Côte et al., 2022). Agroecology revolves around the concept of integral sustainability, which encompasses the sustainable production and resilience of agroecosystems (Gómez et al., 2016).

Agroecology focuses on mobilizing and managing ecological processes for sustainable production, as well as developing interactions between technical, social, environmental, and institutional components of agroecosystems (Côte et al., 2022). Agroecological bioeconomy emphasizes the development of eco-friendly processes for biomass and by-products transformation, supporting crop diversification and valorization for fertilization and protection (Faucon et al., 2023). By applying these principles, agroecological bioeconomy promotes sustainable agricultural practices that optimize natural complementarity between crop species, strengthen virtuous cycles in agricultural production, and create more resilient systems (Côte et al., 2022). These principles aim to integrate ecological, economic, and social dimensions of the food system (Ryan & Peigné, 2017).

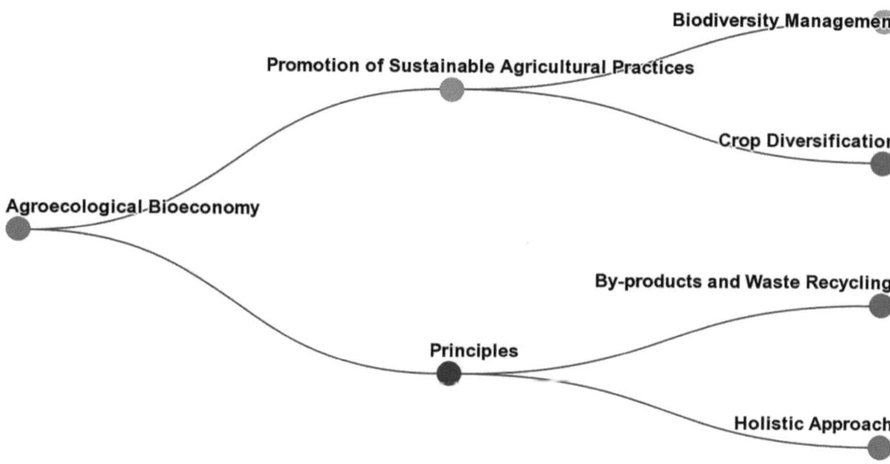

Fig. 10.2 Agroecological bioeconomy principles promote sustainable agricultural practices by integrating ecological, economic, and social dimensions of the food system (Ryan & Peigné, 2017)

Agroecological bioeconomy involves implementing specific management actions on farms to build new agroecological systems (Toffolini et al., 2019). These practices include mobilizing ecological processes; developing interactions between technical, social, environmental, and institutional components of agroecosystems; and promoting individual and collective learning (Côte et al., 2022). It also requires a fundamental redesign of agricultural systems and adaptation to local conditions (Meynard, 2017). In summary, the key principles of agroecological bioeconomy focus on sustainability, resilience, holistic approaches, and scaling up actions. The practices involve mobilizing ecological processes, developing interactions, promoting learning, and redesigning agricultural systems. Agroecological bioeconomy principles promote sustainable agricultural practices by integrating ecological, economic, and social dimensions of the food system (Ryan & Peigné, 2017).

These practices contribute to the reduction of the agricultural carbon footprint and support food security under global change (Faucon et al., 2023). Furthermore, the key principles of agroecological bioeconomy contribute to sustainable agricultural practices by promoting ecological processes, holistic approaches, and scaling up actions. These principles include (Fig. 10.2) mobilization and management of ecological processes: Agroecology focuses on sustainable production and resilience of agroecosystems by utilizing ecological processes. This involves optimizing natural complementarity between crop species to enhance productivity. Development of interactions between components: Agroecology emphasizes the integration of technical, social, environmental, and institutional components of agroecosystems for a holistic approach (Côte et al., 2022). This holistic approach considers the interactions between plants, animals, humans, and the environment within agricultural systems (Ryan & Peigné, 2017). Scaling up of agroecology: Rather than relying solely on increased resources and standardized technical processes, agroecology

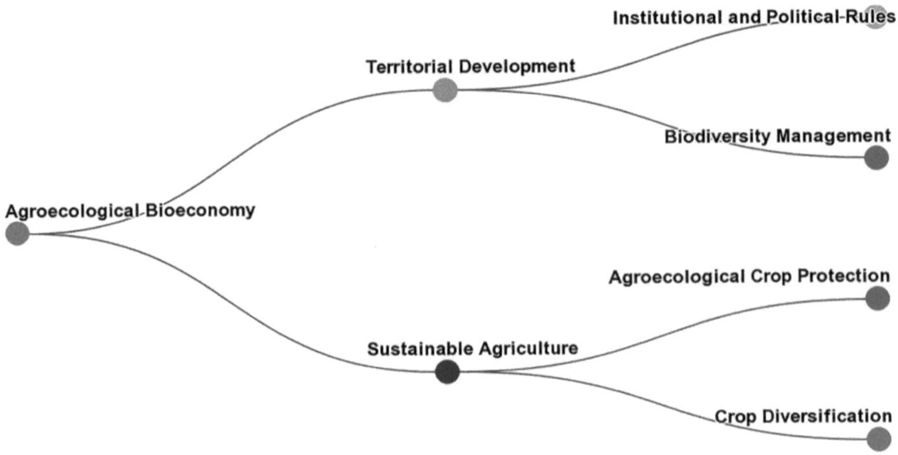

Fig. 10.3 The key principles of agroecological bioeconomy play a crucial role in ensuring long-term sustainability in agricultural practices (Niggli et al., 2023)

emphasizes a plurality of actions and pathways at different organization levels to promote sustainability (Côte et al., 2022). This includes changes in knowledge systems, markets, collaborations, and policies (Niggli et al., 2023).

By following these principles, agroecological bioeconomy contributes to sustainable agricultural practices by reducing trade-offs between productivity and sustainability, promoting biodiversity and ecosystem services, and improving the resilience of food and farming systems (Niggli et al., 2023).

On the other hand, the key principles of agroecological bioeconomy play a crucial role in ensuring long-term sustainability in agricultural practices (Fig. 10.3). Agroecology promotes sustainability by emphasizing the diversity of crops and livestock, fields, farms, and landscapes. It mobilizes biodiversity and ecosystem services for productivity, leading to economically and ecologically resilient production systems (Niggli et al., 2023). Agroecological practices involve the mobilization and management of ecological processes; the development of interactions between technical, social, environmental, and institutional components; and the scaling up of agroecology through various actions and pathways (Côte et al., 2022). The integration of agroecology and bioeconomy creates more resilient agricultural systems. It involves the development of agroecological solutions for the bioeconomy, such as crop diversification and the valorization of by-products for fertilization and protection (Niggli et al., 2023).

By implementing these principles, agricultural systems can achieve sustainability by reducing trade-offs between productivity and environmental impact, promoting resilience, and ensuring the long-term health of ecosystems and communities (Niggli et al., 2023; Côte et al., 2022) (Fig. 10.3).

Similarly, the principles of agroecological bioeconomy can help mitigate the environmental impacts of conventional agricultural practices in several ways (Fig. 10.4). Agroecological practices and eco-friendly processes: Agroecology

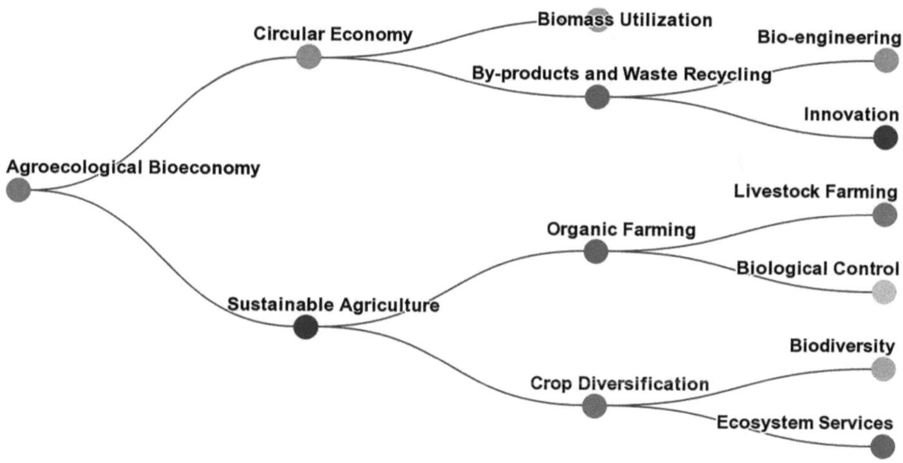

Fig. 10.4 The principles of agroecological bioeconomy can help mitigate the environmental impacts of conventional agricultural practices in several ways (Niggli et al., 2023)

focuses on developing practices that reduce the carbon footprint of agriculture and promote the transformation of biomass and by-products into bio-based products Niggli et al., 2023). These practices can contribute to the reduction of agrochemical use and improve the environmental impact of agricultural production systems (Lampkin et al., 2016). Agroecology emphasizes the importance of crop diversification and the integration of biodiversity in agricultural systems (Hartley, 2018). This can enhance ecosystem services such as pest management, nutrient cycling, and resilience (Hartley, 2018). Biodiversity is key to maintaining healthy ecosystems and reducing the negative environmental impacts of conventional agriculture (Niggli et al., 2023). Agroecological approaches aim to enhance crop-soil productivity in sustainable ways without compromising the environment (Raj et al., 2021). By applying sustainable practices, such as soil management and stress management, agroecosystems can improve soil health and environmental quality (Khanal, 2023). Overall, the principles of agroecological bioeconomy offer a pathway to transform agricultural systems toward sustainability by promoting eco-friendly practices, biodiversity, and sustainable intensification (Faucon et al., 2023; Niggli et al., 2023; Raj et al., 2021; Hartley, 2018; Lampkin et al., 2016).

The adoption of agroecological bioeconomy principles can contribute to the reduction of greenhouse gas (GHG) emissions in the agricultural sector through several mechanisms (Fig. 10.5). This can help reduce N_2O emissions, a potent GHG, by implementing better nitrogen management practices (Van Noordwijk, 2014). By implementing better feeding regimes for ruminants and reducing inundation of rice fields, methane (CH_4) emissions can be reduced (Cardinael et al., 2021a, b; Van Noordwijk, 2014). By reducing CO_2 emissions from peatlands through better management practices, agroecological bioeconomy principles can contribute to GHG reduction (Van Noordwijk, 2014; Verchot et al., 2007). Implementing better waste recycling practices can help reduce GHG emissions in the agricultural sector

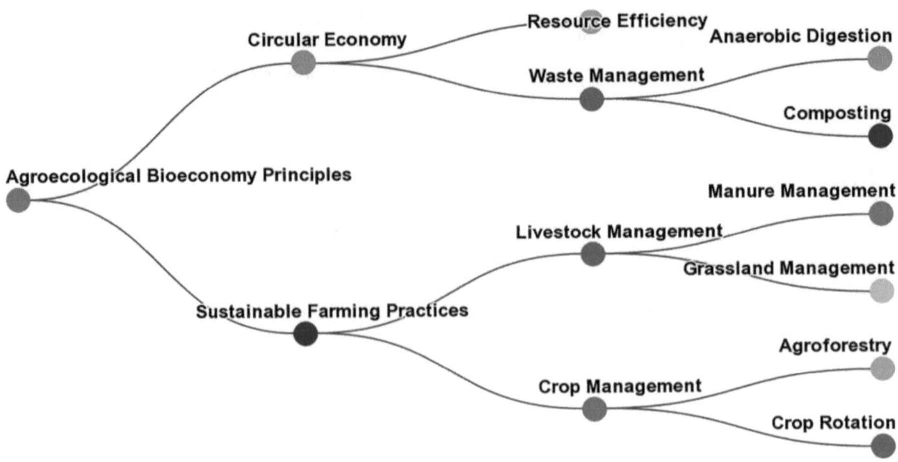

Fig. 10.5 The adoption of agroecological bioeconomy principles can contribute to the reduction of greenhouse gas (GHG) emissions in the agricultural sector (Cardinael et al., 2021a, b; Tang et al., 2018; Van Noordwijk, 2014)

(Van Noordwijk, 2014). Agroforestry and increased carbon storage in soils can help sequester carbon and reduce CO_2 emissions (van Noordwijk et al., 2015; Van Noordwijk, 2014). Shifting toward more sustainable diets and lifestyles can help reduce the overall carbon footprint of the agricultural sector (Van Noordwijk, 2014; Rampin & Rampin, 2021; Paletto et al., 2020).

By adopting these principles, the agricultural sector can contribute to the reduction of GHG emissions and mitigate climate change impacts (Cardinael et al., 2021a, b; Papadopoulou et al., 2021; van Noordwijk et al., 2015; Van Noordwijk, 2014).

10.3 Methodology Issues: Bibliometric Network Analysis

The methodology for the bibliometric analysis involves the retrieval of pertinent scientific publications from established databases like the Web of Science or Scopus. This process includes delineating the sample based on publication year, authors' geographic locations, research fields, thematic journal categorization, or keyword selection (Bugge et al., 2016). The analysis encompasses data processing and recording related to publications, such as citation counts, distribution by country, scientific discipline, authorship, and institutional affiliations (Papavlasopoulos, 2015). Subsequent steps involve identifying research output trends and characteristics at organizational and national levels, gauging the impact and assessment of research, and detecting scientific networks among researchers and areas (Papavlasopoulos, 2015). However, the analysis faces challenges concerning:

10.3.1 Scientific Areas

Comparison of indicators across scientific fields proves intricate due to variations in publication reporting patterns and the time frame of research obsolescence. For instance, scientific publications in cancer research experience high citation rates shortly after publication, while social science publications accrue citations over a more extended period (Papavlasopoulos, 2015).

10.3.2 Time Period Consideration

The time elapsed since publication significantly influences citation counts. To normalize these differences, specific time intervals are established for citation measurements (Glanzel, 2003).

10.3.3 Publication Type Influence

Citation patterns in articles are influenced by subject content, with fields like physics showing lower research activity compared to medicine. Moreover, highly cited research fields tend to receive more citations on average (Moed, 2006; Glanzel, 2003). The analysis employs bibliometric networks visualized through complex nodes and links to convert quantitative data into qualitative insights (Konstantinis et al., 2018). The five fundamental methods for network term correlation include co-authorship, co-occurrence, citation, bibliographic coupling, and co-citation analyses (Papavlasopoulos, 2015).

In this study, the VOSviewer software was utilized for bibliometric network analysis due to its reliability, ease of use, and ability to convert quantitative data into visual representations (Aristovnik et al., 2020; Einecker & Kirby, 2020; Van Eck & Waltman, 2010). Scopus was selected as the primary database for data collection, meeting the criteria of extensive coverage, tools for data extraction, and open access (Tomich et al., 2011).

10.4 Results and Discussion

The study's focal point was the identification of the term "agroecological bioeconomy" within publication titles, abstracts, and keywords. Table 10.1 delineates the prevalence of three key terms ("Agroecology," "Bioeconomy," and "Agroecological Bioeconomy") in academic literature across distinct time spans. "Agroecology" is referenced in 6134 articles spanning from 1953 to 2024, while

Table 10.1 Term frequency in academic publications

Term	Time limitation	Number of articles
Agroecology	1953–2024	6134
Bioeconomy	1979–2024	6015
Agroecological	2015–2024	13
Total		12,162

"Bioeconomy" is present in 6015 articles between 1979 and 2024. Notably, the combined term "Agroecological Bioeconomy" emerges in only 13 articles published from 2015 to 2024 (Faucon et al., 2023). The analysis via VOSviewer culminated in a total count of 12,162 paper publications, emphasizing the intertwining nature of agroecology and bioeconomy within the context of agroecological bioeconomy (Table 10.1).

10.4.1 Keyword Network Analysis

The Agroecology keyword analysis yielded 15,342 unique results, with 1614 keywords meeting the set threshold of appearing at least five times, denoting their higher relevance. Prominent keywords included "Agroecology," "Agriculture," "Sustainability," "Biodiversity," and "Climate Change," with varying occurrences and link strengths. In the Bioeconomy keyword analysis, a total of 21,439 unique keywords were identified, among which 2331 met the five-time threshold. Noteworthy keywords encompassed "Biomass," "Bioeconomy," "Sustainable Development," and "Biofuels," each exhibiting distinct frequencies and link strengths. The analysis of Agroecological Bioeconomy keywords, as depicted in Table 10.2, presented recurring keywords such as "Agroecology," "Agriculture," "Sustainability," "Biomass," "Biodiversity," and "Climate Change," aligning closely with the Agroecology domain (Faucon et al., 2023; Gliessman, 2021; Côte et al., 2022; Francis et al., 2003). Among these, "Biomass," "Sustainability," and "Agriculture" were especially relevant to the concept of Agroecological Bioeconomy, showcasing their recurrent occurrence and link strength within this context (Wertz et al., 2022; Saleem et al., 2020; D'Amato et al., 2020).

Figure 10.6 emphatically reinforces the interconnectedness outlined earlier, underscoring the robust linkage between Agroecological Bioeconomy and pivotal themes like Bioeconomy, Biomass, and Agroecology. This visual representation vividly portrays the strong correlations, reaffirming the profound integration of Agroecology within the realm of Agroecological Bioeconomy. The compelling associations depicted in the figure validate and emphasize the pivotal roles played by Bioeconomy, Biomass, and Agroecology within the context of Agroecological Bioeconomy, further solidifying their integral positions within this domain (Selbonne et al., 2022; Zanetti et al., 2021; Levidow, 2015).

Figure 10.7 illustrates the distribution of sub-areas within Agroecology, Bioeconomy, and Agroecological Bioeconomy domains, depicted as percentages

Table 10.2 The top 10 keywords related to the Agroecological Bioeconomy, ranked by the co-occurrence

Agroecology			Bioeconomy			Agroecological bioeconomy		
Keywords	Co-occurrence	Total link strength	Keywords	Co-occurrence	Total link strength	Keywords	Co-occurrence	Total link strength
Agroecology	2968	1971	Biomass	752	1401	Bioeconomy	7	3
Agriculture	342	668	Bioeconomy	1607	1224	Biomass	3	4
Sustainability	383	598	Sustainable development	565	1088	Agroecology	3	3
Biodiversity	345	571	Nonhuman	391	1063			
Climate change	303	512	Sustainability	520	909			
Alternative agriculture	203	384	biofuels	309	875			
Article	166	383	Biofuel	273	868			
Crop production	187	383	Article	377	820			
human	146	379	Circular bioeconomy	496	695			
Crop yield	178	341	Review	248	649			

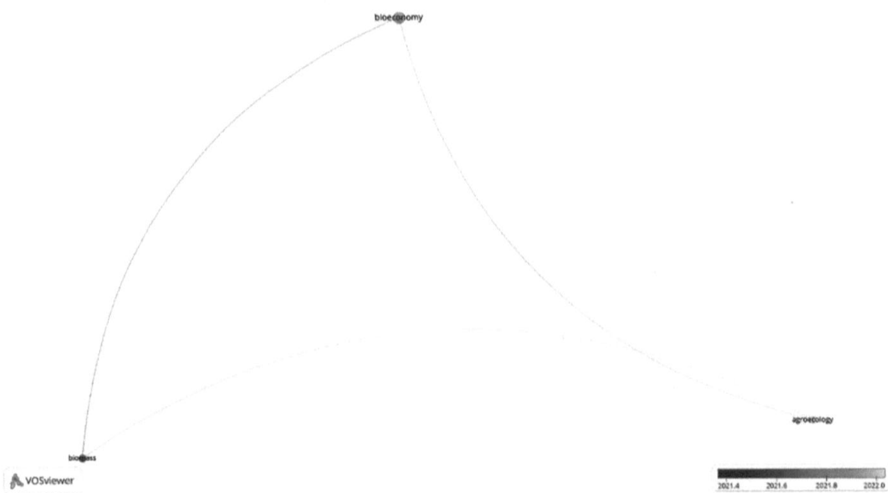

Fig. 10.6 Mapping strong associations: agroecological bioeconomy and its nexus with bioeconomy, biomass, and agroecology (Selbonne et al., 2022; Zanetti et al., 2021; Levidow, 2015)

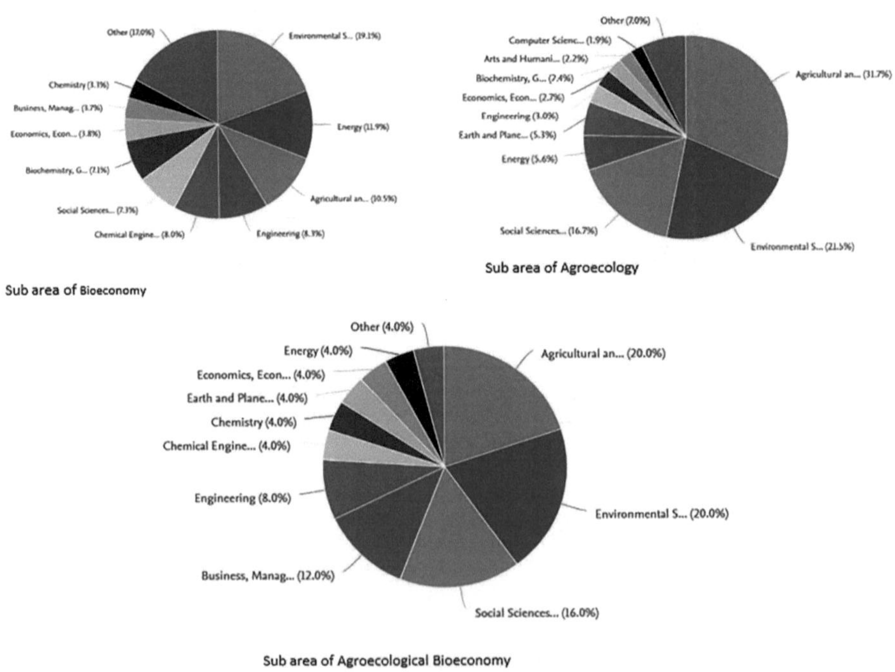

Fig. 10.7 Interdisciplinary distribution of sub-areas in agroecology, bioeconomy, and agroecological bioeconomy (Toruño et al., 2022; Zúniga-González et al., 2022; Dios-Palomares et al., 2015; Simkin & Roychowdhury, 2012)

within each field. Notable distinctions are observed across the disciplines, emphasizing key thematic concentrations and disparities among the three domains. Agroecology exhibits substantial emphasis on Environmental Sciences, Agriculture, Social Sciences, and Earth/Planetary Sciences, while Bioeconomy emphasizes Engineering-related fields, Business Management, and various other disciplines. The synthesis in Agroecological Bioeconomy notably integrates aspects from Environmental Sciences, Agriculture, Social Sciences, and Business Management, highlighting a balanced convergence of disciplines from both Agroecology and Bioeconomy domains (Toruño et al., 2022; Zúniga-González et al., 2022; Dios-Palomares et al., 2015).

10.5 Conclusions

The study extensively analyzed the term "agroecological bioeconomy" within academic literature, focusing on its presence in publication titles, abstracts, and keywords. Data from Table 10.1 revealed that "Agroecology" appears in 6134 articles spanning from 1953 to 2024, while "Bioeconomy" is referenced in 6015 articles from 1979 to 2024. In contrast, the term "Agroecological Bioeconomy" appears in only 13 articles published between 2015 and 2024. This indicates a relatively low focus on the integrated concept of agroecological bioeconomy compared to its constituent terms.

Keyword network analysis showed a significant volume of keywords related to Agroecology and Bioeconomy. For Agroecology, 15,342 unique keywords were identified, with 1,614 meeting the relevance threshold, highlighting key themes such as 'Agroecology,' 'Agriculture,' 'Sustainability,' 'Biodiversity,' and 'Climate Change.' Similarly, the Bioeconomy analysis identified 21,439 unique keywords, with 2,331 meeting the threshold, emphasizing terms like 'Biomass,' 'Bioeconomy,' 'Sustainable Development,' and 'Biofuels'.

The analysis of Agroecological Bioeconomy keywords (Table 10.2) indicated that "Biomass," "Sustainability," and "Agriculture" were particularly relevant, showing strong co-occurrence and link strength. This reinforces the interconnectedness of Agroecology and Bioeconomy within the Agroecological Bioeconomy framework. Figures 10.6 and 10.7 further illustrate the robust relationships between these domains, with Fig. 10.6 demonstrating strong associations with Bioeconomy, Biomass, and Agroecology. Figure 10.7 underscores the interdisciplinary nature of Agroecological Bioeconomy, integrating Environmental Sciences, Agriculture, Social Sciences, and Business Management.

In conclusion, while "Agroecological Bioeconomy" is emerging as a specialized field, it remains less prevalent in academic literature compared to its broader components. The findings emphasize the need for increased research focus on this integrated concept and its practical applications, highlighting its potential to bridge gaps between Agroecology and Bioeconomy and foster interdisciplinary advancements.

Acknowledgments Not applicable.

Author Contributions Conceptualization, ZG.-C.A. and E.L.; methodology, ZG.-CA., E.L., K. M., and F.C.; software, ZG.-C.A.; validation, E.L., K.M., and F.C.; data curation, C.-I.P.; writing—original draft preparation, C.-I.P.; writing—review and editing, C.-I.P., E.L., K.M., and F.C.; visualization, C.-I.P.; supervision, E.L. All authors have read and agreed to the published version of the manuscript.

Institutional Review Board Statement Not applicable.

Informed Consent Statement Not applicable.

Data Availability Statement Not applicable.

Conflicts of Interest The authors declare no conflict of interest.

Funding This research received no external funding.

References

Aristovnik, A., Ravšelj, D., & Umek, L. (2020). A bibliometric analysis of COVID-19 across science and social science research land-scape. *Sustainability, 2020*(12), 9132. https://doi.org/10.3390/su12219132

Backhouse, M., Lühmann, M., & Tittor, A. (2022). Global inequalities in the bioeconomy: thinking continuity and change in view of the global soy complex. *Sustainability, 14*(9), 5481. https://doi.org/10.3390/su14095481

Bugge, M. M., Hansen, T., & Klitkou, A. (2016). What is the bioeconomy? A review of the literature. *Sustainability, 2016*(8), 691. https://doi.org/10.3390/su8070691

Cardinael, R., Cadisch, G., Gosme, M., Oelbermann, M., & Van Noordwijk, M. (2021a). Climate change mitigation and adaptation in agriculture: Why agroforestry should be part of the solution. *Agriculture, Ecosystems & Environment, 319*, 107555. https://doi.org/10.1016/j.agee.2021.107555

Cardinael, R., Cadisch, G., Gosme, M., Oelbermann, M., & Van Noordwijk, M. (2021b). Climate change mitigation and adaptation in agriculture: Why agroforestry should be part of the solution, *Agriculture, Ecosystems & Environment, 319*, 107555, ISSN 0167-8809. https://doi.org/10.1016/j.agee.2021.107555. (https://www.sciencedirect.com/science/article/pii/S0167880921002590).

Côte, F. X., Rapidel, B., Sourisseau, J. M., Affholder, F., Andrieu, N., Bessou, C., Patrick, C., Jean-Philippe, D., Faure, G., Hainzelin, E., Malezieux, E., Poirier-Magona, E., Roudier, P., Scopel, E., TixierP, Toillier, A., & Perret, S. (2022). Levers for the agroecological transition of tropical agriculture. *Agronomy for Sustainable Development, 42*(4), 67. https://doi.org/10.1007/s13593-022-00799-z

D'Amato, D., Bartkowski, B., & Droste, N. (2020). Reviewing the interface of bioeconomy and ecosystem service research. *Ambio, 49*, 1878–1896. https://doi.org/10.1007/s13280-020-01374-0

Dios-Palomares, R. Alcaide, D. Diz, J., Jurado, M. Prieto, A. Morantes, M., & Zúñiga, C. A. (2015). Análisis de la eficiencia de sistemas agropecuarios en América latina y el Caribe mediante la incorporación de aspectos ambientales. *Revista Científica, XXV*(1), 43–50. Redalyc, https://www.redalyc.org/articulo.oa?id=95934122007

Einecker, R., & Kirby, A. (2020). Climate change: A bibliometric study of adaptation, mitigation and resilience. *Sustainability, 2020*(12), 6935. https://doi.org/10.3390/su12176935

Faucon, M. P., Aussenac, T., Debref, R., Firmin, S., Houben, D., Marraccini, E., Sauvée, L., Trinsoutrot-Gattin, I., & Gloaguen, R. (2023). Combining agroecology and bioeconomy to meet the societal challenges of agriculture. *Plant and Soil, 492*(1), 61–78. https://doi.org/10.1007/s11104-023-06294-y

Feleke, S., Cole, S. M., Sekabira, H., Djouaka, R., & Manyong, V. (2021). Circular bioeconomy research for development in sub-Saharan Africa: innovations, gaps, and actions. *Sustainability, 13*(4), 1926. https://doi.org/10.3390/su13041926

Foguesatto, C. R., Artuzo, F. D., de Oliveira, L., & Leal, Â. R. (2017). Agenda de pesquisa da bioeconomia: Um estudo no campo do agronegócio. *Agenda, 38*(04).

Francis, C., Lieblein, G., Gliessman, S., Breland, T. A., Creamer, N., Harwood, R., Salomonsson, L., Helenius, J., Rickerl, D., Salvador, R., Wiedenhoeft, M., Simmons, S., Allen, P., Altieri, M., Flora, C., & Poincelot, R. (2003). Agroecology: The ecology of food systems. *Journal of Sustainable Agriculture, 22*(3), 99–118. https://doi.org/10.1300/J064v22n03_10

Gil, J., Alter, E., La Rota, M. J., Tello, E., Galletto, V., Padró, R., Martínez, T., Darnay, S., & Marull, J. (2022). Towards an agroecological transition in the Mediterranean: A bioeconomic assessment of viticulture farming. *Journal Of Cleaner Production, 380*, 134999. https://doi.org/10.1016/j.jclepro.2022.134999

Giraldo, O. F., & Rosset, P. M. (2018). Agroecology as a territory in dispute: Between institutionality and social movements. *The Journal of Peasant Studies, 45*(3), 545–564. https://doi.org/10.1080/03066150.2017.1353496

Glanzel, W. (2003). BIBLIOMETRICS as a research field a course on theory and application of bibliometric indicators. Available online: https://www.cin.ufpe.br/~ajhol/futuro/references/01%23_Bibliometrics_Module_KUL_BIBLIOMETRICS%20AS%20A%20RESEARCH%20FIELD.pdf. Accessed 02 Jan 2024.

Gliessman, S. R. (2021). *Package price agroecology: The ecology of sustainable food systems*. CRC Press.

Gómez, L. F., Ríos Osorio, L. A., & Eschenhagen Durán, M. L. (2016). Key concepts of agroecology science. a systematic review.

Hartley, S. E. (2018). Agroecological approaches to sustainable intensification. In *Sustainable food and agriculture: an integrated approach* (pp. 179–184). https://doi.org/10.1016/B978-0-12-812134-4.00013-3

Isaac, M. E., Isakson, S. R., Dale, B., Levkoe, C. Z., Hargreaves, S. K., Méndez, V. E., WittmanH, Hammelman, C., Langill, J. C., Martin, A. R., Nesol, E., Ekers, M., Borden, K. A., Gagliardi, S., Buchanan, S., Archibald, S., & Gálvez Ciani, A. (2018). Agroecology in Canada: towards an integration of agroecological practice, movement, and science. *Sustainability, 10*(9), 3299. https://doi.org/10.3390/su10093299

Khanal, N. (2023). Sustainable agriculture and cultivation practices. In *Sustainable food science - A comprehensive approach* (Vol. 1(4)), pp. 30–50. https://doi.org/10.1016/B978-0-12-823960-5.00080-9

Konstantinis, A., Rozakis, S., Maria, E. A., & Shu, K. (2018). A definition of bioeconomy through the bibliometric networks of the scientific literature. *AgBioforum, 2018*(21), 64–85.

Lampkin, N. H., Smith, J., & Smith, L. G. (2016). Agroecology and organic farming as approaches to reducing the environmental impacts of agricultural chemicals. *Issues in Environmental Science and Technology*, 94–113. https://doi.org/10.1039/9781782626916-00094

Levidow, L. (2015). European transitions towards a corporate-environmental food regime: Agroecological incorporation or contestation? *Journal of Rural Studies, 40*, 76–89. https://doi.org/10.1016/j.jrurstud.2015.06.001

Lewandowski, I. (2018). *Bioeconomy: Shaping the transition to a sustainable, biobased economy* (p. 356). Springer nature. https://doi.org/10.1007/978-3-319-68152-8

Lima, M. G. B. (2022). Just transition towards a bioeconomy: Four dimensions in Brazil, India and Indonesia. *Forest Policy And Economics, 136*, 102684. https://doi.org/10.1016/j.forpol.2021.102684

Meadows, D. H., Meadows, D. L., Randers, J., & Behrens III, W. W. (1972). *The limits to growth.* Universe Books, 211 p.

Meynard, J. M. (2017). Agroecology, a new relationship to knowledge and innovation. *OCL-Oilseeds and Fats, Crops and Lipids, 24*(3).

Moed, H. F. (2006). Citation analysis in research evaluation. Springer Science & Business Media; Part of the book series: Information Science and Knowledge Management (ISKM, volume 9). https://doi.org/10.1007/1-4020-3714-7 XIV, 348.

Niggli, U., Sonnevelt, M., & Kummer, S. (2023). Pathways to advance agroecology for a successful transformation to sustainable food systems. In *Science and innovations for food systems transformation* (pp. 341–359). Springer International Publishing. https://doi.org/10.1007/978-3-031-15703-5_18

Paletto, A., Biancolillo, I., Bersier, J., Keller, M., & Romagnoli, M. A. (2020). Literature review on forest bioeconomy with a biblio-metric network analysis. *Journal of Forest Science, 66*, 265–279. https://doi.org/10.17221/75/2020-JFS

Papadopoulou, C. I., Loizou, E., Melfou, K., & Chatzitheodoridis, F. (2021). The knowledge based agricultural bioeconomy: A bibliometric network analysis. *Energies, 14*(20), 6823. https://doi.org/10.3390/en14206823

Papavlasopoulos, S. (2015). Scientific reporting and measurement tools in Bibliometrics; Publisher Kallipos: .

Pimbert, M. P. (2018). Global status of agroecology: A perspective on current practices, potential and challenges. *Economic and Political Weekly, 53*(41), 52–57.

Pyka, A., Cardellini, G., van Meijl, H., & Verkerk, P. J. (2022). Modelling the bioeconomy: Emerging approaches to address policy needs. *Journal of Cleaner Production, 330*, 129801. https://doi.org/10.1016/j.jclepro.2021.129801

Rampin, R., & Rampin, V. (2021). Taguette: open-source qualitative data analysis. *Journal of Open Source Software, 6*(68), 3522. https://doi.org/10.21105/joss.03522

Raj, A., Jhariya, M. K., Khan, N., Banerjee, A., Paikra, P. R., Meena, R. S., & Kumar, S. (2021). Intensification for agroecosystem services. In *Sustainable Intensification for Agroecosystem Services and Management* (pp. 197–228). https://doi.org/10.1007/978-981-16-3207-5_7

Ryan, M. R., & Peigné, J. (2017). Applying agroecological principles for regenerating soils. In *Agroecological practices for sustainable agriculture: Principles, applications, and making the transition* (pp. 53–84). https://doi.org/10.1142/9781786343062_0003

Saleem, F. Khattak, A. Rehman, S. U., & Ashiq, M. (2020). Bibliometric analysis of green marketing research from 1977 to 2020. *Publications 2021, 9*, 1. https://doi.org/10.3390/publications9010001

Schwarz, G. Vanni, F. Miller, D. Helin, J. Pražan, J. Albanito, F.. Fratila, M. Galioto,F. Gava, O. Irvine, K. Landert, J. Linares Quero, A. Mayer, A. Monteleone, D. Muller, A. Röös, E. Smyrniotopoulou, A. Vicente, A. Vlahos, G.& Zīlāns, A. (2022). Exploring sustainability implications of transitions to agroecology: a transdisciplinary perspective. EuroChoices, 21(3), 37–47. https://doi.org/10.1111/1746-692X.12377.

Selbonne, S., Guindé, L., Belmadani, A., Bonine, C., Causeret, F. L., Duval, M., & Blazy, J. M. (2022). Designing scenarios for upscaling climate-smart agriculture on a small tropical Island. *Agricultural Systems, 199*, 103408. https://doi.org/10.1016/j.agsy.2022.103408

Simkin, M. V., & Roychowdhury, V. P. (2012). Theory of citing. In *handbook of optimization in complex networks* (pp. 463–505). Springer. https://doi.org/10.1007/978-1-4614-0754-6_16

Tang, M., Liao, H., Wan, Z., Herrera-Viedma, E., & Rosen, M. A. (2018). Ten years of sustainability (2009 to 2018): A biblio-metric overview. *Sustainability, 10*, 1655. https://doi.org/10.3390/su10051655

Toffolini, Q., Cardona, A., Casagrande, M., Dedieu, B., Girard, N., & Ollion, E. (2019). Agroecology as farmers' situated ways of acting: a conceptual framework. *Agroecology and Sustainable Food Systems, 43*(5), 514–545. https://doi.org/10.1080/21683565.2018.1514677

Tomich, T. P., Brodt, S., Ferris, H., Galt, R., Horwath, W. R., Kebreab, E., Johan, H. J., Leveau, D., Liptzin, M., Lubell, P., Merel, P., Michelmore, R., Rosenstock, T., Scow, K., Six, J., Williams, N., & Yang, L. (2011). Agroecology: a review from a global-change perspective. *Annual Review of Environment and Resources, 36*, 193–222. https://doi.org/10.1146/annurev-environ-012110-121302

Toruño, P. J., Zuniga-Gonzalez, C. A., Castellón, J. D., Hernández-Rueda, M. J., & Gutierrez-Espinoza, E. I. (2022). Identificación de los Senderos productivos de la Bioeconomía en Universidades del CNU y sector agropecuario. *Revista Iberoamericana de Bioeconomía y Cambio Climático, 8*(16), 1929–1946. https://doi.org/10.5377/ribcc.v8i16.15016

Van Eck, N. J., & Waltman, L. (2010). Software survey: VOSviewer, a computer program for bibliometric mapping. *Scientometrics, 84*, 523–538. https://doi.org/10.1007/s11192-009-0146-3

Van Noordwijk, M. (2014). Climate change: agricultural mitigation. Encyclopedia of agriculture and food systems (pp. 220–231), 1 January 2014. ISBN 978-008093139-5. https://doi.org/10.1016/B978-0-444-52512-3.00002-4

van Noordwijk, M., Minang, P. A., & Hairiah, K. (2015). Swidden transitions: In an era of climate-change debate. In *Shifting cultivation and environmental change* (pp. 261–280). Routledge.

Verchot, L. V., Van Noordwijk, M., Kandji, S., Tomich, T., Ong, C., Albrecht, A., Mackensen, J., Bantilan, C., Anupama, K. V., & Palm, C. (2007). Climate change: linking adaptation and mitigation through agroforestry. *Mitigation and Adaptation Strategies for Global Change, 12*, 901–918. https://doi.org/10.1007/s11027-007-9105-6

Viaggi, D. (2018). *The bioeconomy: delivering sustainable green growth*. CABI. https://doi.org/10.1079/9781786392756.0000

von Braun, J. (2018). Bioeconomy-the global trend and its implications for sustainability and food security. *Global Food Security, 19*, 81–83. https://doi.org/10.1016/j.gfs.2018.10.003

Von Cossel, M., Wagner, M., Lask, J., Magenau, E., Bauerle, A., Von Cossel, V., Warrach-Sagi, K., Elbersen, B., Staritsky, I., Van Eupen, M., Iqbal, Y., Jablonowski, N. D., Happe, S., Fernando, A. L., Scordia, D., Cosentino, S. L., Wulfmehyer, V., Lewandowski, I., & Winkler, B. (2019). Prospects of bioenergy cropping systems for a more social-ecologically sound bioeconomy. *Agronomy, 9*(10), 605. https://doi.org/10.3390/agronomy9100605

Wertz, J. L., Mengal, P., & Perez, S. (2022). *Biomass in the bioeconomy: Focus on the EU and US*. CRC Press. https://doi.org/10.1201/9781003308454

Wezel, A., Herren, B. G., Kerr, R. B., Barrios, E., Gonçalves, A. L. R., & Sinclair, F. (2020). Agroecological principles and elements and their implications for transitioning to sustainable food systems. A review. *Agronomy for Sustainable Development, 40*, 1–13. https://doi.org/10.1007/s13593-020-00646-z

Zanetti, F., Alberghini, B. M., Jeromela, A., Grahovac, N., Rajković, D., Kiprovski, B., & Monti, A. (2021). Camelina, an ancient oilseed crop actively contributing to the rural renaissance in Europe. A review. *Agronomy for Sustainable Development, 41*, 1–18. https://doi.org/10.1007/s13593-020-00663-y

Zúniga-González, C. A., López, M. R., Icabaceta, J. L., Vivas-Viachica, E. A., & Blanco-Orozco, N. V. (2022). Epistemology of bioeconomy. *Iberoamerican Journal of Bioeconomy and Climate Change, 8*(15), 1786–1796. https://doi.org/10.5377/ribcc.v8i15.13986

Chapter 11
Exergetic Scope of Bioeconomics: The Sustainable Energy Transition

Salvador Peniche-Camps and Sergio Gabriel Ceballos Pérez

Abstract This chapter explores the nature of the energy transition toward renewable technologies in Mexico. It focuses on the relevance of the threat posed by climate change, an emergency primarily derived from the effects of fossil fuel combustion. It argues that, under existing technological conditions in Mexico, the gap between the energy density of fossil fuels and the potential energy of renewable technologies—the exergy of alternative technologies—in addition to the energy policy shift of the last seven years, makes the transition to a predominance of renewable energy an unlikely scenario even in these times. Given the energy requirements demanded by population and economic growth, greater adjustments and investments are required, especially in infrastructure and transition technology. Such is the case of bioenergy, which has significant untapped potential but can be used in circular models for agroindustry, urban areas, etc.

Keywords Technophilia · Ecological limits · Energy density · Potential energy · Biophysical economics

11.1 Introduction

Due to the primary factor of the climate crisis being the generation of greenhouse gases (such as carbon dioxide, produced by hydrocarbon combustion, which accounts for about 65% of the total), obtaining the necessary energy to meet the needs of contemporary society through "renewable" energies has become a priority (Al-Yasiri & Géczi, 2021).

S. Peniche-Camps (✉)
Department of Economics, University of Guadalajara, Zapopan, Jalisco, México
e-mail: speniche@cucea.udg.mx

S. G. Ceballos Pérez
Dirección de Investigación, Innovación y Posgrado, Universidad Politécnica de Pachuca, Zempoala, Hidalgo, México
e-mail: sgceballospe@secihti.mx

S. G. Ceballos Pérez et al. (eds.), *Sustainability in Latin America*,
https://doi.org/10.1007/978-3-032-09787-3_11

This study is important as it critically evaluates the challenges of transitioning to renewable energy in a country highly dependent on hydrocarbons. By examining energy efficiency through the concept of Energy Return of Investment (EROI) and using Mexico as a case study, the research contributes to the field of bioeconomy by providing empirical data and analytical frameworks that highlight the complexities of sustainable energy transitions. This analysis offers valuable insights for policy development aimed at achieving realistic, long-term energy goals in resource-limited settings.

In light of this, the chapter seeks to resolve the following question: Given the energy demand, the level of technological development, and the existing difference between the energy density of fossil fuels and the potential offered by alternative energy sources, is the transition strategy viable?

To answer the research question posed, we will focus on installed energy capacity and analyze the energy market in Mexico in the context of the depletion of hydrocarbon reserves and its economic situation.

The study's theoretical foundation is biophysical economics, and its foundational concept, EROI (Energy Return on Investment), involves evaluating the opportunity cost of exploiting hydrocarbon reserves.

The work begins with the Sect. 11.2 which outlines the study's theoretical basis. In the Sect. 11.3, we provide an analysis of the energy market in the country to offer an empirical explanation of the energy transition context. In the Sect. 11.4, we explore quantitative scenarios related to the possibility of meeting energy demand in the absence of fossil fuels in the context of economic growth.

The methodology employs an interdisciplinary approach centered on biophysical economics to assess the viability of the energy transition in Mexico. It uses EROI (Energy Return on Investment) to measure the energy potential of various sources and establishes quantitative scenarios that compare the capacity of renewable energies against the depletion of hydrocarbons. A historical-comparative analysis is included to contrast past resource availability with current limitations, providing an empirical framework for understanding the implications of the energy transition in the country.

11.2 Bioeconomy Versus Biophysical Economics and the Limits of Development

The purpose of the bioeconomy is to utilize biological resources and biomass to create a nature-based economy that is not dependent on fossil fuels (Hodson de Jaramillo, 2018). However, the definition of bioeconomy is unclear and, on the contrary, ambiguous. The term bioeconomy is often confused with other terms such as biophysical economics, ecological economics, environmental economics, green economics, and life economics, among others (Nifatova et al., 2024). The bioeconomy differs from the previous ones in four key aspects:

- Conservation and valuation of biodiversity and biomass: This involves protecting and responsibly using biological resources, including agrobiodiversity, to ensure the resilience and sustainability of agri-food systems.
- Incorporation of science, technology, and innovation: The aim is to use science, technology, and innovation, including ancestral knowledge, to develop innovative solutions in the bioeconomy.
- Value addition and circularity: The creation of products and services with greater added value is promoted, as well as the implementation of circular economic models that allow for the efficient use of resources and waste reduction.
- Sustainability and social equity: The aim is to ensure that the bioeconomy is environmentally sustainable and benefits all people, including rural producers and local communities.

Besides, the foundational premise of biophysical economics' theoretical proposal is to prioritize the preservation of ecological balances over the imperatives of economic efficiency in the development strategy. By definition, this recognizes the principle of *degrowth* proposed by heterodox economic theory (Latouche, 2008; Martínez-Alier, 2009).

> In general, degrowth challenges the hegemony of economic growth and calls for a democratically redistributive reduction of production and consumption in industrialized countries to achieve environmental sustainability, social justice, and well-being (Demaria et al., 2013). Degrowth is often associated with the idea that more minor can be beautiful. However, the emphasis should be on more than just less and different. "In a degrowth society, everything will be different: activities, forms, and uses of energy, relationships, gender roles, time allocations between paid and unpaid work, and relationships with the non-human world". (Demaria, 2021: 4)

Thus, more than a theory, "Degrowth" constitutes an ideology that focuses on the dematerialization of production, patterns of natural resource utilization, and, notably, the decolonization of economic thought. It is a reaction against the dogma of economic growth detached from the environmental context, which, since the beginnings of capitalism with the emergence of banks and later with the drive of the Industrial Revolution, was presented as the indisputable recipe for achieving social progress.

Today, the teaching and practice of economics as a science embody the idea of growth and the dominant idea of sustainability. Indeed, growth is considered a sine qua noncondition in the environmental policies proposed by international agencies. The environmental interpretation of the Kuznets curve is the theoretical justification for this stance, suggesting that an increase in Gross Domestic Product reduces environmental destruction beyond the curve's inflection point (Catalán, 2014).

Empirical verification of such a theoretical principle is difficult to consider if, within the context of analysis, we include the externalization of environmental costs in the international economy, particularly the increase in ecological debt incurred by industrialized countries with poor countries. Evidence suggests that developed countries with relatively acceptable environmental standards exist due to poor countries' role as pollution sinks or sources of raw materials in the global market.

In this regard, biophysical economics, as articulated by one of its founders, Kenneth Boulding (1996), posits a straightforward principle: the impossibility of infinite growth in a finite environment. This axiom, which lies at the epicenter of the current sustainability debate, is expressed in the theoretical discourse on technophilia and environmentalism: the dichotomy between strong sustainability (recognizing the biophysical limits of development) and weak sustainability (considering the possibility of remediation).

This theoretical conflict over the potential of technology to resolve the contradiction between the imperative of growth and the depletion and destruction of nature constitutes the frontier between opposing positions on sustainability, growth, and degrowth.

Regarding the energy issue, Biophysical Economics posits that fossil fuels are the actual vector of growth and "social progress." According to this theory, the development of capitalism is explained by the availability of energy resources, their easy access, and relatively low costs rather than by economic policies or various strategies derived from economic theories. The use of fuels, which enable development, has not been determined through the propositions of economic rationality. On the contrary, it is the access to cheap energy sources, particularly oil, that explains the development of the theories that give rise to the development of capitalism.

The above refers to the theory of the "empty world" and the "full world" as explanations for development policies. In an empty world characterized by abundant natural resources, the overexploitation of nature did not play a significant role in development strategies. Only with the collapse of ecosystems have biophysical limits started to feature on the development agenda.

Biophysical Economics explains the environmental effects of hydrocarbon overexploitation through the laws of thermodynamics. According to this perspective, CO2 emissions are the logical result of anthropogenic carbon cycle disruption. Therefore, this phenomenon represents a chemical reaction that will accelerate its effects on the environment by the law of entropy.

However, within the parameters of orthodox economic theory, the challenge is to reconcile destruction with the imperatives of growth without questioning the model. This is demonstrated by the fact that the Sustainable Development Goals (SDGs) agenda does not provide any analytical consideration or recommendations regarding the need to reconsider the growth imperative.

> Due to the exponential economic growth since World War II, we now live in a "full world," yet we still behave as if it were empty, with ample space and resources for an indefinite future. The foundational assumptions of neoclassical economics, developed in the "empty world," no longer hold, as the cumulative burden of the human species is reaching (or, in some cases, exceeding) the limits of nature at local, regional, and planetary levels. The prevailing obsession with economic growth leads us toward ecological collapse, sacrificing the foundation of our well-being and survival. To reverse this ominous trajectory, we must transition to a steady-state economy focused on qualitative development rather than quantitative growth and recognize the interdependence between the human economy and the ecosphere. Developing policies and institutions for a steady-state economy will require us to reevaluate the purpose and goals of the economy. (Daly, 2015)

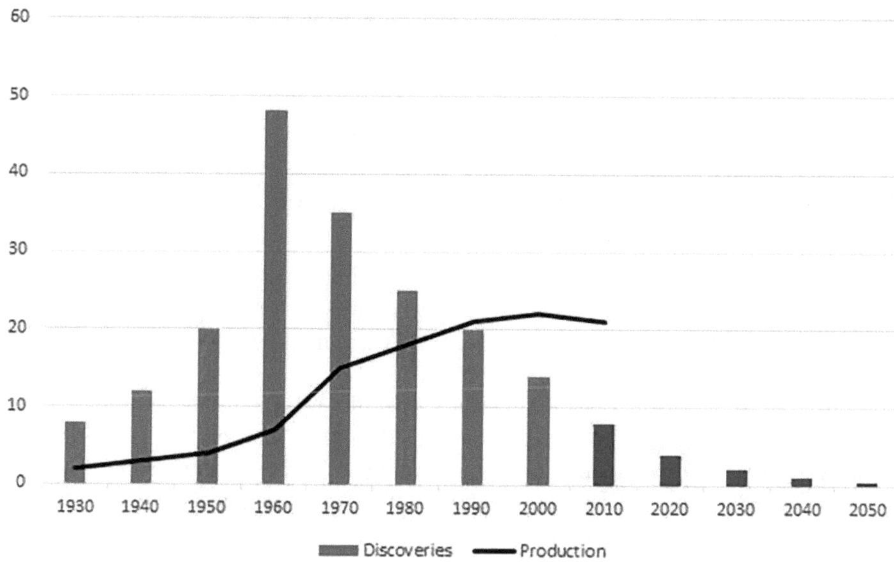

Fig. 11.1 Huber peak of oil (billions of barrels). (Source: Prepared by the author based on Leloux (2011))

Biophysical economics draws on energy return on investment (EROI). The idea behind this principle is that, with the depletion of energy reserves and increasing extraction difficulty, the energy costs per unit of resource extraction rise until viability is lost. The above suggests an economic-biophysical update to Ricardo's theory of land rent—which focuses on the distribution of land value in the market based on fertility—by pointing instead to the accelerated and permanent loss of opportunity cost in the energy generation model due to resource depletion.

This concept allows us to advance an interpretation of the disarray in the oil price system and the intensification of current social and geopolitical conflicts. EROI provides a robust framework for understanding economic scarcity in the case of energy resources and constitutes an empirical application of one of the essential principles in natural resource economics, namely, that the viability of resource exploitation does not depend on physical assets but on opportunity costs (Hotelling, 2001).

To illustrate how EROI functions within the context of interpreting the concept of degrowth, let us examine Fig. 11.1.

The graph in Fig. 11.1 shows the global Hubbert peak of oil. This methodological tool, which graphically describes the evolution of energy reserves, is helpful for directly highlighting the fundamental proposition of Biophysical Economics. The graph illustrates, on the one hand, the temporal coincidence of global economic growth in the last century with the increase in the rate of oil field discoveries and, on the other hand, the trend toward a decrease in this rate with the decline in oil production. The question that arises from the graph is: What will be the source that replaces the energy supply that will be vacant with the decline in oil reserves under conditions of economic growth? The discussion proposed in the essay the reader has in hand derives from the answer to this question: either demand is reduced (degrowth theory) or supply is increased (energy transition theory).

11.3 The Decline of Mexican Energy Model

This section aims to provide the energy sector's context in relation to the proposal for a transition toward the production and consumption of renewable energies.

Let us start with oil production and exports. A downward trend has been observed since 2004 (see Fig. 11.2).

However, it is worth questioning oil's specific weight in the overall energy structure and the prevailing trend. Figure 11.3 illustrates that total oil consumption, along with the other two hydrocarbons, maintains an upward proportional trend.

The sector's structure and trend align with global trends, where the energy industry's prevalence and growth are concentrated in the extraction of fossil fuels.

In Mexico, the evolution of the oil sector relates to the discovery of one of the most significant oil fields in world history: the Cantarell oil reserve. Discovered in 1966 and operational by 1972, the field once contributed 45% of global production at its peak (Romo, 2023).

Unfortunately, the field's exploitation using incorrect technologies, such as nitrogen injection, led to its collapse. Although it still provides energy along with other oil fields in the country, much of its potential has been lost or remains underground, significantly affecting Mexico's energy potential (see Fig. 11.4).

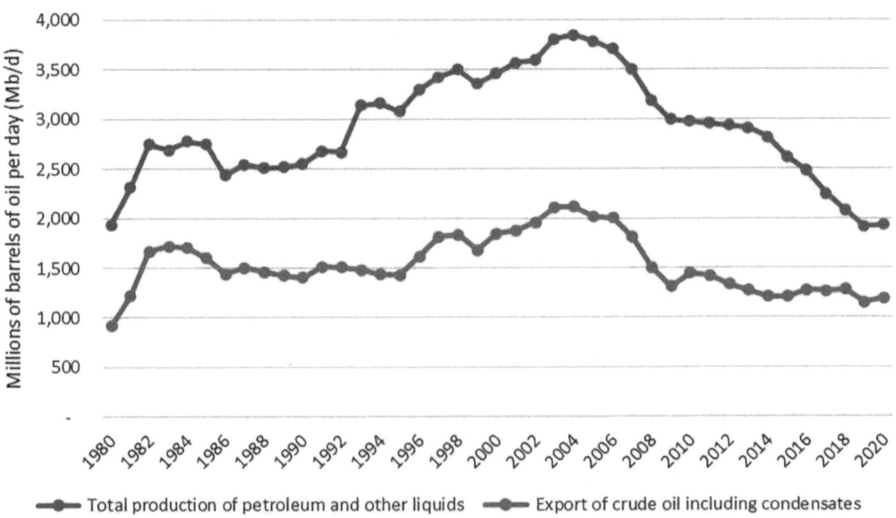

Fig. 11.2 Oil production and exports in Mexico, 1980–2020. (Source: Energy Information Administration (2023b))

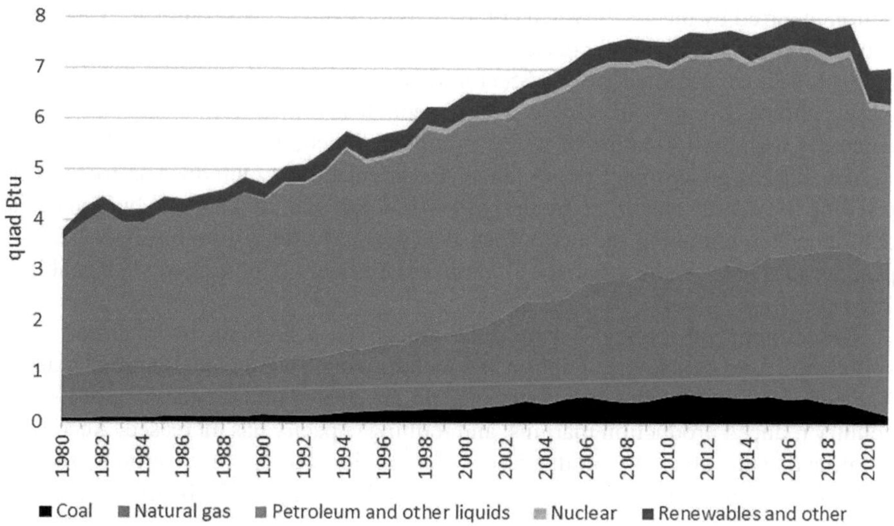

Fig. 11.3 Energy consumption in Mexico, 1980–2021. (Source: Energy Energy Information Administration (2023a))

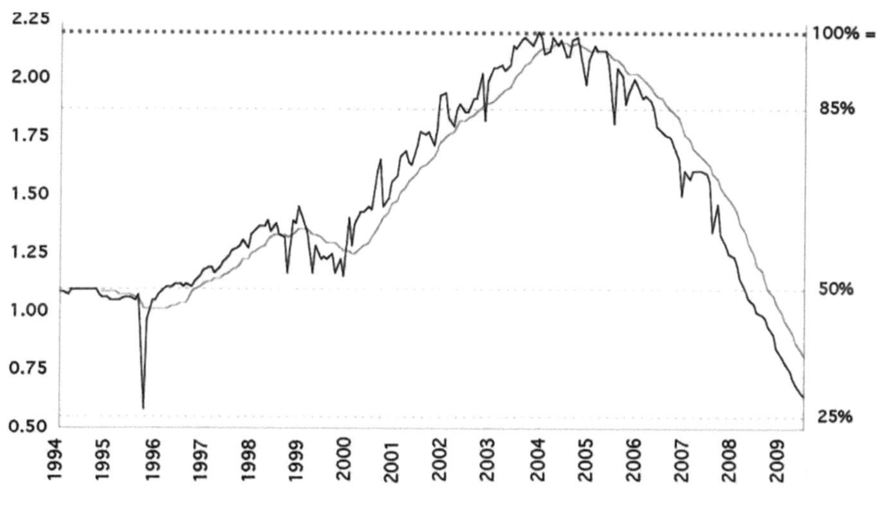

Fig. 11.4 Cantarell field production (Mb/d). (Source: Lajous (2015))

It is fair to say that Mexico's economic success at the end of the last century was fundamentally due to access to cheap oil from the Cantarell field, which was positioned in the international market as Mexico's main export product during those years.

The analysis of oil production, exports, and prices during the golden era of the oil industry in Mexico raises questions about the fate of the resources generated by the Mexican oil company (PEMEX) and the justification for the enormous debt of $67 billion incurred by the state-owned company (Gobierno de la República, 2024). The country's most significant increase in oil exports occurred between 2000 and 2015. In 2011, Mexico obtained around $56 billion due to international prices of $117 per barrel, absent revenues in social development. It is worth noting that this period coincides with the formation of power groups in the country (México, 2024).

The current government policy under President López Obrador has focused on rescuing the energy sector, particularly PEMEX. Despite the financial burden of the debt incurred, removing the tax burden that severely affected its balance sheets in past decades has created a stable economic environment (Gobierno de la República, 2024).

The current oil policy has a dual objective: achieving hydrocarbon self-sufficiency for national consumption and withdrawing from the international market. The national energy project does not include the sale of oil abroad. As a result, the country reduced production margins, and refining capacity has increased. According to official data, daily oil production is 3.4 million barrels, and refining reaches 794,000. The goal for 2026 is to reach 1.778 million barrels (Gobierno de México, 2024).

Given the above, the question arises about the federal government's stance on the energy transition strategy. The exploitation of fossil fuels has not only gone unchallenged but has an explicit objective of ensuring access to fossil fuels for the next decades.

How can this strategy be understood in light of the strengthening of the oil sector in leading countries and the onset of Hubbert peaks on a global scale? Why are there no signs of transition elsewhere in the world? We will address these questions in the following section.

11.4 The Accounts of Energy Transition (the Case of Solar Energy)

The objective of the third section is to provide elements for discussing the viability of transitioning from an energy system based on the exploitation of fossil fuels to a model of renewable energy generation and consumption. According to official sources, fossil fuels represent 72.9% of the total primary energy supply, while renewable sources together contribute only 33.1% (Fig. 11.5). The structure of the energy matrix has become more diversified; however, the energy transition is worrying considering the expectations of economic and demographic growth.

Despite the growth of renewable energy over the last 5 years, the goals committed to in the General Law on Climate Change derived from the Paris Agreement have not been met; however, they are close to achieving them (35% renewable energy by 2024). Renewable energy generation has been increasing slowly but progressively, and the diversity of these sources, especially bioenergy, has grown (SENER, 2022).

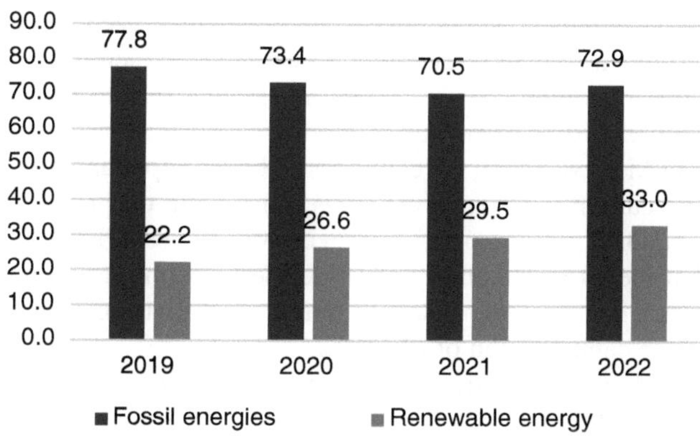

Fig. 11.5 Proportion of fossil energy vs. renewable energy (percentage). (Source: SENER, 2024)

Table 11.1 Generation of renewable energy by type of technology (gigawatts-hour)

Energy type	2019	2020	2021	2022	Average (percentages)
Total hydroelectric	23,602.40	26,817.00	34,717.20	35,558.90	41.329
Geothermal	5060.70	4574.60	4242.90	4412.70	6.491
Total wind power	16,726.90	19,702.90	21,074.90	20,528.80	27.056
Total photovoltaic	9964.30	15,835.60	20,194.90	20,342.00	22.378
Total bioenergy	1866.50	2206.50	1595.60	2141.30	2.746
Sugarcane bagasse	1476.30	1583.20	1374.10	1918.20	2.215
Biogas	241.2	526.7	176.1	153.8	0.396
Biogas-distributed generation				38.9	0.012
Sanitary landfill	110.9	67.4	16.2	0	0.078
Black liquor	38.1	26.4	24.8	23.7	0.041
Biomass	0	2.8	4.3	3.4	0.003
Biomass-distributed generation				3.3	0.001
Total	57,220.80	69,136.60	81,825.40	82,983.60	100.000

Source: SENER (2024)

Table 11.1 shows the total renewable energy sources. The main source is hydroelectric, followed by wind and photovoltaic, which represent just over 90% of renewable energy, while geothermal energy represents 6.5% and bioenergy 2.7%, which is composed of bioenergy from sugarcane bagasse, biogas, landfills, black liquor, and biomass. The transition to renewable energy involves a large number of factors. Beyond policies and laws, it requires public and private investment, technology, and training.

It must also be recognized that relying on a single type of energy represents a risk to the system. Mexico's needs and conditions are another variable to consider. For example, in the case of remote rural communities, rural communities with thriving agro-industrial development, small towns, and large cities, each space requires and has access to different sources of energy. Likewise, due to their geographic location or environmental conditions, some communities are likely to have wind power, others solar energy, hydroelectric power, etc.

Therefore, in Mexico's case, oil continues to be an important part of the energy mix despite the current environmental crisis, fluctuating oil prices, and the bankruptcy of its national oil company, Petróleos Mexicanos (PEMEX). To this we should add that the current administration decided not only to reduce support for renewable energy but, in some cases, to inhibit its growth in favor of fossil fuel extraction and the capitalization of PEMEX.

In the case of biomass energy, it represents only 0.004% of the total combined energy use with distributed generation. However, according to the International Energy Agency (IEA), biomass energy has vast potential that has not yet been tapped, both due to its availability and low costs. Mexico currently has 33 bioenergy plants with a total installed capacity of 408 megawatts (MW). Bioenergy generation accounted for 0.63% of the total, representing a share equivalent to 2147.3 GWh of generation in 2022 (SENER, 2023). In addition, regarding this type of energy source (Bioenergy), the states with the highest share are as follows: Veracruz with 192 MW; Oaxaca with 50 MW; San Luis Potosí with 49 MW; and Hidalgo with 37 MW and an electric power generation of 7989.2 GWh in 2078 to 2747.3 GWh in 2022.

From a thermodynamic-social perspective, the problem we face is as follows: As a society, we are undergoing a biophysical and chemical transformation of the environment, which involves the anthropogenic disruption of the carbon cycle through the overexploitation of fossil fuels. In this sense, climate change appears as another stage in the planet's long and tumultuous geological history, characterized by the complexity and interdependence of environmental processes.

The planet's geological history shows millions of years of transformative processes that culminated in the geological epoch known as the Holocene, which provided the conditions for the emergence of human civilization. However, the current epoch, the Anthropocene (named for the human impact as the primary transformative force), is characterized by disrupting the geological conditions necessary for life. Major transformative cataclysms, such as the Great Oxidation or the Snowball Earth period, have been recurrent in the planet's geological history. Indeed, today, we understand that the conditions for the emergence and development of life are infrequent due to their relatively brief temporal span: It involves the disruption of the carbon cycle or the relocation of carbon into the atmosphere.

With "social progress" and the use of highly entropic technologies (those with high "energy waste"), human society has harnessed the kinetic work stored in carbon molecules, which was sequestered underground during centuries of the planet's geological history. In the modern era, the work once performed by enslaved people in ancient civilizations has been transferred to machines powered by fossil fuels. This process has allowed most of humanity to live in ways that only emperors could before the era of modern energy: Joules are the modern enslaved people.

However, as with any socio-environmental process, trade-offs are the norm. The massive use of fossil fuels for social development has consequences, which we term climate change. In response to this scenario, the remediation and adaptation strategy has been the so-called energy transition, which, in thermodynamic terms, could be defined as the social effort to reverse the disruption of the carbon cycle on Earth.

Several considerations compel us to reassess the expectations of the transition, including technological, economic, and socio-environmental factors, which appear widely in the scientific literature (Hall, 2018; Alpert, 2012; Klitgaard & Kral, 2012). According to Ferrari and Ocampo Téllez (2019), the energy differential between oil and solar technology is 5 to 1, meaning that, due to various biophysical, economic, social, and environmental factors, solar energy only generates one-fifth of the installed power capacity, so performing an accelerated transition could cause an energy deficit in the system.

"Low energy density and the intermittency of renewables imply that for every MW of fossil fuel, we require at least five MW of renewable energy, which entails a much larger infrastructure volume. One MW from a combined cycle power plant can deliver seven GWh annually. In contrast, wind energy delivers only about two GWh yearly, and solar energy delivers just one GWh yearly" (Ferrari & Ocampo Téllez, 2019, p. 31).

Considering energy consumption in Mexico and the potential for work it represents, considering its diversification, it could contribute to its sustainable use. Certainly, the analysis of consumption and its diversification (intermediate and final consumption and the process of transforming potential into kinetic energy, for example) adds complexity to the analysis, which is not the objective of this chapter. However, it is also necessary to point out that the transition to renewable energy is not general but can occur by sector, according to their needs, environmental and economic costs, among others. A case in point is the bioenergy produced by landfills, which in the case of Pachuca Hidalgo generates a significant amount of energy to supply public lighting (Ceballos et al., 2022).

In 2022, Mexico's national energy consumption reached 1,513,333 GWh, of which 70.1% came from fossil fuels, such as gasoline, diesel, liquefied gas, natural gas, coal, and coke, while the remaining 29.9% came from renewable energy (SENER, 2023). This means that meeting this need, given that, with current technology, renewable energy can only generate a third of the total, would require a significant increase in infrastructure and investment.

Such a rough calculation does not take into account, for example, expectations for economic growth or population growth, factors that would increase energy demand over time. The former is relevant for several reasons: the increase in energy demand exhibits exponential growth, and the depletion of reserves is not linear.

Consequently, the tipping point of the energy transition, when the descending curve of access to cheap fossil fuels intersects with the increase in renewable energy supply, will not meet the supply limit. Instead, we can foresee scarcity to worsen to the extent that it imposes the need to reassess development projects and their linkage to social progress. In this scenario, the crucial question is not whether reviewing the economic growth and consumption strategy is prudent but when is the right moment to address this dilemma.

Given this more or less evident perspective, energy policy in Mexico and worldwide focuses on efficiently ensuring the existing hydrocarbon reserves, i.e., slowing down the collapse. Mexico's perspective on the issue is reflected in its energy plans, which promise accessible oil for the next 30 years. Such a perspective is attainable in countries with access to reserves or deposits. As Ahmed (2017) asserts, the intensification of social violence in nation-states and international conflicts increasingly reflects the effects of energy scarcity.

11.5 Conclusions

The crossroads humanity faces regarding its future in the coming decades deserves superlative attention. Despite the scientific evidence pointing to a grim scenario, economic policies and public awareness do not react with the seriousness the situation warrants. The issue at hand is the future impact of the fossil fuel crisis expressed among other things in rising concerns about fossil fuel combustion and climate change. There is a perception that this is just another crisis in human history that can be resolved, like previous ones, through innovation and creativity.

However, evidence shows that the scenario of absolute fossil fuel scarcity, which has been the driving force of contemporary society and the source of high living standards, food, and health on a global scale, has begun to manifest.

The dominant approach reflects technological optimism, faith in science, and technology's ability to resolve the dilemma through renewable energies, which theoretically could replace fossil fuels and reverse the disruption of the geological carbon cycle. As the laws of thermodynamics state, energy can change forms, and the challenge for human civilization is to transform the technological base established during the Industrial Revolution—primarily the exploitation of fossil fuels, especially oil—toward other abundant and renewable sources.

Nevertheless, the complexity of the problem requires a more detailed approach to understand the limits and possibilities of this transition. Empirical evidence raises serious questions about the technological feasibility of the transition, given the current technological and environmental conditions. This situation forces us to address a central issue in the current debate on energy, namely, whether to continue with the strategy of economic growth and the accelerated exploitation of resources and energy or to reconsider social development patterns and align them with the clear signals of exhaustion and disruption from bio-physical processes.

The lessons are clear in the case examined in the work at hand, specifically Mexico's energy policy.

1. A defined strategy is to secure access to cheap oil for domestic social development. Awareness of the problem of resource depletion has led authorities to shift from past decades' oil production and exportation strategy to a policy focused on safeguarding reserves for current and future national needs.

2. The rescue of the federal oil agency is a sine qua non to executing this strategy. However, the proposed strategy lacks clarity regarding the economic-financial, technological, and geopolitical context that such a goal requires, especially in the face of the accelerated exacerbation of aggressive domination and plunder policies by dominant global actors.

References

Ahmed, N. (2017). *Failing states, collapsing systems. Biophysical triggers of political violence.* Springer.

Alpert, J. (2012, July 23). *How much degrowth is enough? Rv9* [Video]. YouTube. https://www.youtube.com/watch?v=fcQYI4yo8mM&t=5s&ab_channel=JackAlpert

Al-Yasiri, Q., & Géczi, G. (2021). *Global warming potential: Causes and consequences.* ResearchGate. https://www.researchgate.net/publication/354294388_Global_Warming_Potential_Causes_and_Consequences

Boulding, K. (1996). The economics of the coming spaceship earth. In H. Jarrett (Ed.), *Environmental quality in a growing economy* (pp. 3–14). Resources for the Future/Johns Hopkins University Press. http://arachnid.biosci.utexas.edu/courses/THOC/Readings/Boulding_SpaceshipEarth.pdf

Catalán, H. (2014). Curva ambiental de Kuznets: Implicaciones para un crecimiento sustentable. *Economía Informa,* (389), 19–37. https://www.elsevier.es/es-revista-economia-informa-114-pdf-S0185084914721723

Ceballos S, Brambila J, & Pérez V. (2022). Residuos sólidos urbanos y economía circular en Pachuca, Hidalgo, México. *Acta Universitaria, 32*, e3437. Epub 24 de octubre de 2022. https://doi.org/10.15174/au.2022.3437.

Daly, H. (2015). Economics for a full world. *Great Transition Initiative.* https://greattransition.org/images/Daly-Economics-Full-World.pdf.

Demaria, F. (2021). Decrecimiento: una propuesta para fomentar una transformación socioecológica profundamente radical. *Oikonomics,* (16). https://oikonomics.uoc.edu/divulgacio/oikonomics/es/numero16/dossier/fdemaria.html

Energy Information Administration, EIA. (2023a). Independent statistics and analysis. Mexico, primary energy.

Energy Information Administration, EIA. (2023b). *Independent statistics and analysis.* Mexico, petroleum and other liquids. https://www.eia.gov/international/data/country/MEX

Ferrari, L., & Ocampo Téllez, E. (2019). Tendencias globales en energía y perspectiva de México. En J. L. Calva Futuro de la energía en México (pp. 17-44). Consejo Nacional de Universitarios.

Gobierno de México (2024). Versión estenográfica. https://www.gob.mx/presidencia/articulos/version-estenografica-conferencia-de-prensa-del-presidente-andres-manuel-lopez-obrador-del-4-de-enero-de-2024

Hall, C. (2018). *Energy and the wealth of nations. An introduction to biophysical economics.* Springer.

Hodson de Jaramillo, E. (2018). Bioeconomía: el futuro sostenible. *Revista de la Academia Colombiana de Ciencias Exactas, Físicas y Naturales, 42*(164), 188–201. https://doi.org/10.18257/raccefyn.650

Hotelling, H. (2001). Economía de los recursos agotables. *Gestión y Ambiente, 4*(1), 87–109. https://revistas.unal.edu.co/index.php/gestion/article/view/88829/75443

Klitgaard, K., & Kral, L. (2012). Ecological economics, degrowth, and institutional change. *Ecological Economics, 84*, 247–253. https://doi.org/10.1016/j.ecolecon.2011.11.008

Lajous, A. (2015). *Declinación de las reservas.* OCE. https://energia.org.mx/el-ocaso-de-cantarell/

Latouche, S. (2008). *La apuesta por el decrecimiento. ¿Cómo salir del imaginario dominante?.* Icaria.

Leloux, J. (2011). El pico del petróleo. *Entenderlo. Asumirlo. Superarlo. ResearchGate*. 10.13140/ 2.1.3148.2243.

Martínez-Alier, J. (2009). Hacia un decrecimiento sostenible en las economías ricas. *Revista de Economía Crítica, 8*, 121–137. https://revistaeconomiacritica.org/index.php/rec/article/view/42 6/409

Nifatova, O., Danko, Y., Petrychuk, S., & Romanenko, V. (2024). Modern bioeconomy measurement in the green economy paradigm: Four pillars of alternative bioeconomy. *Sustainability, 16*(22), 9612. https://doi.org/10.3390/su16229612

Romo, D. (2023). El campo petrolero Cantarell y la economía mexicana. *Problemas Del Desarrollo. Revista Latinoamericana De Economia, 46*(183). https://doi.org/10.1016/j.rpd. 2015.10.007

SENER. (2022). *Balance nacional de energía.* [Formato de presentación] https://base.energia.gob. mx/BNE/BalanceNacionalDeEnerg%C3%ADa2022.pdf

SENER. (2023). *Atlas Nacional de Zonas con Alto Potencial de Energías Limpias*. Gobierno de México.

SENER. (2024). *Reporte de avance de energías limpias 2024*. Gobierno de México,

Index

S. G. Ceballos Pérez et al. (eds.), *Sustainability in Latin America*,
https://doi.org/10.1007/978-3-032-09787-3

MIX
Papier aus verantwortungsvollen Quellen
Paper from responsible sources
FSC® C105338

If you have any concerns about our products,
you can contact us on
ProductSafety@springernature.com

In case Publisher is established outside the EU,
the EU authorized representative is:
Springer Nature Customer Service Center GmbH
Europaplatz 3, 69115 Heidelberg, Germany

Printed by Libri Plureos GmbH
in Hamburg, Germany